W9-BZZ-314

LONELY PLANET'S

Armchair Explorer

Contents

Clockwise from top left: New Zealand singer-songwriter Bic Runga; be part of your *Tokyo Story*; Gabriel García Márquez in Cartagena; Prague, home to Franz Kafka and Milan Kundera.

ULF ANDERSEN / GETTY IMAGES; MARK READ / LONELY PLANET

Introduction

Right: the library of Trinity College Dublin. This is a great literary city, home to not only James Joyce but also Flann O'Brien, Samuel Beckett, Maeve Binchy and Jonathan Swift.

'Toto, I have a feeling we're not in Kansas any more.'

That first inkling you have of being away from home is when you turn the dial on a radio and hear music and songs that you don't recognise. Or you switch on a television and see scenes from a movie that was never shown in your home-town cinema. Nothing grounds a traveller more quickly than being immersed in the culture – the books, films and music – of another place. Whether it's the sensory overload of a Bollywood movie in Mumbai or singing along to a French pop song (by law, one third of the music played by radio stations in France has to be in the French language), experiencing the local culture is a great way to get in tune with a new country's general approach to life. Books too, whether carried as creased paperbacks or stored in an app, are windows into other worlds.

But you don't have to pack a suitcase to read, listen and view the books, music and movies from other countries: you can do it from the comfort of your own armchair. And it's a great way of researching a country before you go or simply enjoying one of the finest qualities of travel – gaining an understanding of a faraway place and its people and being enriched by their creative expression – from home.

And that's what this book aims to do: it provides lists of books to read, films to watch and songs to listen to for more than 120 countries around the world. You may not get the chance to travel to each in your lifetime but you can visit them whenever you choose.

Tour Russia with Dostoevsky as a guide, join James Joyce on a mind-bending day out in Dublin, or live in 19th-century Australia as outlaw Ned Kelly via writer Peter Carey. Tune into chilled-out Ethiopian jazz or bop along to Icelandic pop with Björk. Sit down to sample effortlessly stylish French New Wave cinema or be spirited away by Japanese anime. The world can come to you, if you know where to start.

This book was compiled by Lonely Planet writers, who have collectively spent countless years on the road in these countries, Shazam-ing local radio and picking up reading tips from locals. With their guidance you can enjoy a taste of anywhere from Argentina to Zimbabwe, whether you're planning your next trip or exploring the world from your armchair.

nn
mm
ll
kk
ii
hh
gg
ff

EUR

OPE

Austria

Read List

Radetzky March
by Joseph Roth (1932, trans 1933)
Roth's acclaimed novel is about a great family's rise and fall from the 1850s to the 1910s, from war glory to gambling, as cracks widen in the Austro-Hungarian Empire.

The Man Without Qualities
by Robert Musil (1930-1943, trans 1953)
This hugely influential modernist epic was 20 years in the making. It's comfortably over 1000 pages yet was never completed. At its heart is Ulrich, a mathematician in the grip of existential doubt, in a society in slow decline.

The World of Yesterday
by Stefan Zweig (1942, trans 1943)
Zweig, who was Jewish, wrote this moving memoir in exile; his descriptions of life in early 20th-century Vienna take in school, sexuality, social change and the rise of fascism.

The Piano Teacher
by Elfriede Jelinek (1983, trans 1988)
A piano teacher seeks to escape her difficult mother via a masochistic affair with one of her students. This stark novel of obsession became a Michael Haneke movie.

The Dog King
by Christoph Ransmayr (1995, trans 1997)
Ransmayr's vivid alt-history describes three characters attempting to survive in a wild wasteland created by the Morgenthau Plan – an American scheme to destroy the postwar industry of Germany and Austria that was discussed but never carried out.

Watch List

Sissi
(1955; drama; directed by Ernst Marischka)
Princess Elisabeth of Bavaria falls in love with the young emperor of Austria, but he's engaged to her sister. One of the most successful German-language films ever, it's a parade of magnificent costumes and ornate settings.

The Sound of Music
(1965; musical; directed by Robert Wise)
Based (with much added drama) on the memoirs of Maria von Trapp, this smash hit musical features the gorgeous buildings, meadows and lakes in and around Salzburg.

Funny Games
(1997; drama; directed by Michael Haneke)
Haneke's provocative arthouse films have won multiple awards. This, his most notorious work, tells of a family whose lakeside holiday home is terrorised by two strangers, as reality blurs and boundaries are crossed.

Museum Hours
(2012; drama; directed by Jem Cohen)
In this subtle and playful drama, an Austrian museum guard helps a Canadian tourist and they end up exploring Vienna, art history and life's big questions.

The Dreamed Ones
(2016; drama; directed by Ruth Beckermann)
Two young actors meet in a recording studio to chat, smoke and read the correspondence between the poets Paul Celan and Ingeborg Bachmann, who were lovers and creative rivals in postwar Austria.

Vienna's alternative culture is expressed in its music scene and in architecture such as the Hundertwasser House.

Austria produced and nurtured an astonishing array of composers in the 18th and 19th centuries. Joseph Haydn and Wolfgang Amadeus Mozart were followed by Franz Schubert and Johannes Brahms. The Salzburg-born Mozart wrote some 626 pieces, including *Don Giovanni* (1787), *Così Fan Tutte* (1790) and *The Magic Flute* (1791).

Schlager is a European genre of catchy, accessible pop with sentimental lyrics. Udo Jürgens scored big *schlager* hits in the 1950s, before broadening his style into chanson, disco and mainstream pop. A Eurovision Song Contest winner, he filled arenas until his death in 2014.

Playlist

The Marriage of Figaro
Wolfgang Amadeus Mozart
Genre: Classical

Cello Concerto No 2 in D Major
Joseph Haydn
Genre: Classical

Rock Me Amadeus
Falco
Genre: Pop

Seite an Seite
Christina Stürmer
Genre: Pop

Notturno for Strings and Harp
Arnold Schönberg
Genre: Classical

Higher
Edenbridge
Genre: Metal

Griechischer Wein
Udo Jürgens
Genre: Schlager

Booty Swing
Parov Stelar
Genre: Electric swing

Anton Aus Tirol
Anton & DJ Ötzi
Genre: Schlager

High Noon
Kruder & Dorfmeister
Genre: Electronica

Belarus

Read List

Voices from Chernobyl
by Svetlana Alexievich (1997)
The first Belarusian to win the Nobel Prize (in 2015) writes gripping, personal accounts about great crises — including this unflinching portrait of death and heroism in the catastrophic nuclear reactor accident of 1987.

King Stakh's Wild Hunt
by Uladzimir Karatkevich (1964, trans 2012)
This macabre detective novel set in the late 19th century provides fascinating insight into imperial Russia. It's rich with Karatkevich's typical flourishes, including poetic language and elaborately drawn characters.

Paranoia
by Victor Martinovich (2010, trans 2013)
Part thriller and part dystopian fiction, *Paranoia* is a remarkable tale of love and fear set in a thinly veiled version of Minsk. The book was banned in Belarus, and Martinovich now lives in exile.

Burning Lights
by Bella Chagall (1946)
Evoking a lost world, Chagall writes movingly of her Jewish upbringing in pre-war Vitebsk. Beautiful illustrations by her famous husband Marc Chagall accompany the text.

The Dead Feel No Pain
by Vasil Bykau (1965, trans 2010)
One of the few WWII novels written from the perspective of a Soviet soldier. Bykau paints an evocative portrait of life and death on the harrowing Eastern Front battlefields.

Watch List

Viva Belarus!
(2012; drama; directed by Krzysztof Łukaszewicz)
This film shows the harsh repression faced by those brave enough to speak out against the Lukashenko dictatorship. It's inspired by the true story of Belarusian activist Franak Viachorka.

Ivan & Abraham
(1993; drama; directed by Yolande Zauberman)
Set in the 1930s, two friends from opposite backgrounds — one Christian, one Jewish — flee the confines of village life against the impending conflict of Nazi invasion.

Come and See
(1985; drama; directed by Elem Klimov)
One of history's most powerful war movies depicts, in nightmarish and surreal detail, a 13-year-old boy from a Nazi-occupied village who joins the Belarusian partisans.

Dangerous Acts Starring the Unstable Elements of Belarus
(2013; documentary; directed by Madeleine Sackler)
A look at the actions of the Belarus Free Theatre, which staged politically charged productions around the globe, speaking out against tyranny to bring attention to state-sponsored repression.

Franz + Polina
(2006; drama; directed by Mikhail Segal)
A surprising wartime love story between two people who can't speak the same language, but are willing to make the ultimate sacrifice for one another.

The National Library of Belarus keeps a copy of every book published in the country in this extraordinary building.

The famed song *Try Carapachi* became an unofficial anthem for thousands of protestors who packed the streets after Belarus' 2020 presidential election, which many considered fraudulent. Guitar-wielding Pit Pawlaw, one of NRM's frontmen, faced down the riot police and led the crowds in song.

Going strong since 1999, Stary Olsa stay fresh by performing modern rock songs on medieval instruments. Flutes, Belarusian bagpipes and lutes are put to surprisingly effective use in a version of Black Sabbath's *Iron Man*.

Playlist

Cassiopeia
Zlata Mai
Genre: Pop

Try Carapachi
NRM
Genre: Rock

Mechta
IOWA
Genre: Indie-rock

Optimist
Max Korzh
Genre: Hip-hop

Peremen
Gryaz
Genre: Rock

Ginseng Woman
Mustelide
Genre: Electro-pop

Pesnya Schastlivykh Ludey
Anna Sharkunova
Genre: Pop

Dva Kruki
Stary Olsa
Genre: Folk

Sudno
Molchat Doma
Genre: Post-punk

August
Intelligency
Genre: Techno-blues

Belgium

Read List

Pietr the Latvian
by Georges Simenon (1931, trans 1933)
This is the first outing for pipe-smoking detective Maigret. Here he's after a mysterious con man, who is found dead on a train – but also appears to be alive and well, in various locations around France.

My Little War
by Louis Paul Boon (1947, trans 2010)
Boon fought in WWII, was captured and spent the rest of the war in a POW camp and his home village. He wrote notes on the people he met, and turned them into this compassionate collection of stories.

Wonder
by Hugo Claus (1962, trans 2009)
Poet, playwright, novelist and painter Claus was one of the 20th century's leading Belgian writers. *Wonder* follows a teacher's journey to a village of Nazi sympathizers.

Speechless
by Tom Lanoye (2009, trans 2016)
Lanoye shook up Belgian literature as an *enfant terrible* in the 1980s; this contemplative work sees him look back on his actress mother's life, and Belgium's recent history, as her health declines.

Thirty Days
by Annelies Verbeke (2015, trans 2016)
Verbeke's funny, perceptive novel follows a charismatic Senegalese migrant who moves in and out of the lives of people across his Flanders community.

Watch List

Man Bites Dog
(1992; drama; directed by Rémy Belvaux, André Bonzel & Benoît Poelvoorde)
This controversial tale of a serial killer who roams a Belgian town with a camera crew in tow is a grisly litany of robbery and murder – and a bleakly funny satire.

Rosetta
(1999; drama; directed by Jean-Pierre Dardenne & Luc Dardenne)
The Dardenne brothers specialise in gritty films about poverty and crime, and have had a huge impact on Belgian cinema; this story of a teenager's struggle to find work won the Palme d'Or at Cannes.

JCVD
(2008; drama; directed by Mabrouk el Mechri)
In this postmodern tangle Jean-Claude Van Damme plays against type: as a dejected version of himself, who returns to Brussels and gets swept up in a bank robbery.

The Misfortunates
(2009; comedy drama; directed by Felix Van Groeningen)
A boy grows up with his layabout father and uncles in a film that mixes comedy with a more serious exploration of how childhood relationships shape us as adults.

Bullhead
(2011; drama; directed by Michaël R Roskam)
Matthias Schoenaerts bristles with emotion as a cattle farmer with a dark secret in this gripping, Flanders-set thriller about organised crime and the beef industry.

The ever-dapper musician Plastic Bertrand was born in Brussels and christened Roger François Jouret.

Clubs in Brussels and Ghent were instrumental in popularising electronic music in the 1980s, with curfew-free venues playing European body music and new beat, which fused disco, rock and early house. *Pump up the Jam* emerged in 1989 as dance music first went mainstream, and was a hit across Europe and North America.

The Belgian-born Romani guitarist Django Reinhardt's joyful fingerwork pioneered 'gypsy jazz' in the 1930s and '40s. He toured widely, and his Quintette du Hot Club de France was one of Paris' leading bands during a golden age for European jazz.

Playlist

Ça Plane Pour Moi
Plastic Bertrand
Genre: Punk

Pump up the Jam
Technotronic
Genre: House

Yeux Disent
Lomepal
Genre: Hip-hop

Suds & Soda
dEUS
Genre: Indie

Symphony in D Minor
César Franck
Genre: Classical

Ne Me Quitte Pas
Jacques Brel
Genre: Chanson

Le Banana Split
Lio
Genre: Pop

Minor Swing
Django Reinhardt
Genre: Jazz

Je T'aime
Lara Fabian
Genre: Pop

Is It Always Binary?
Soulwax
Genre: Electronica

IN DEPTH

Tintin

from BELGIUM

© JACQUES PAVLOVSKY / GETTY IMAGES

Left to right: Hergé working on another episode of Tintin's adventures; the Hergé Museum in Brussels, designed by Christian de Portzamparc, opened in 2009; a Belgian airliner in Tintin-inspired livery.

© PJOAOMARTINSARC / GETTY IMAGES

Hergé span of some of the 20th century's greatest yarns. His *Adventures of Tintin* have sold over 200 million copies in 70 languages, and transformed comic books.

Born Georges Remi in 1907 to a Walloon father and a Flemish mother, he showed an interest in art from an early age, and got his first break via a strip in Belgium's boy scouting magazine. In 1929, he created Tintin, a bequiffed young reporter who defeated dictators and gangsters with quick wits, courage and the odd knockout punch. Over the next 50 years, Hergé gave his straightforward hero a vivid supporting cast – including whisky-loving Captain Haddock, eccentric Professor Calculus and resourceful fox terrier Snowy.

Hergé's colourful panels are packed with affectionate jokes and impossible drama. His 24 Tintin books take in Inca gold, underwater treasure, cruel slave-traders, bumptious insurance salesmen and bizarre meteorites. Hergé has his critics, and his earliest titles leaned heavily on colonial stereotypes. Yet his impact has been enormous, his *ligne claire* ('clear line') style influencing Andy Warhol and renowned graphic novelist Chris Ware, and his work celebrated in Belgium's inventive Musée Hergé. His cheerful, action-packed tales move at a rattling pace, but can be returned to again and again – many readers find Tintin is a friend for life

Bosnia & Hercegovina

Read List

The Bridge on the Drina
by Ivo Andrić (1945, trans 1959)
No book better captures Bosnian life under Ottoman rule than this magnificent intergenerational epic set around the famous bridge in Višegrad by Bosnia-born Andrić, a Nobel laureate.

Bosnian Chronicle
by Ivo Andrić (1945, trans 1958)
So brilliant he gets two spots on this list, Andrić set this novel in 19th-century Travnik, the former seat of the Ottoman viziers and the town of his birth.

Death and the Dervish
by Meša Selimović (1966, trans 1996)
Another classic of Bosnian literature, this book follows the sheik of a Dervish community (a mystical branch of Islam) as he struggles to free his brother from an Ottoman prison.

Sarajevo Marlboro
by Miljenko Jergović (1994, trans 1997)
This moving collection of short stories gives well-observed and sometimes humorous vignettes of daily life during the siege of Sarajevo (1992–1996), written by one of its survivors.

Shards
by Ismet Prcic (2011)
Set during the Bosnian War, this semi-autobiographical novel sees the young protagonist escape the embattled city of Tuzla and flee to the US where he struggles to fit in.

Watch List

The Battle of Neretva
(1969; action; directed by Veljko Bulajić)
This Oscar-winning epic highlights a major battle in Bosnia during WWII between the Yugoslav Partisans on one side and the Germans, Italians and Chetniks (Serbian nationalists) on the other.

When Father Was Away on Business
(1985; drama; directed by Emir Kusturica)
Set in the paranoid environment of the 1950s, Sena tells her son that his father's away on business when he's really a political prisoner in a labour camp.

No Man's Land
(2001; drama; directed by Danis Tanović)
A winner of an Oscar and Golden Globe for Best Foreign Language Film, *No Man's Land* focuses on a Bosniak and a Serbian soldier trapped together in a trench between their two opposing sides.

Grbavica
(2006; drama; directed by Jasmila Žbanić)
Set in the Grbavica suburb of postwar Sarajevo, a young girl finally discovers the truth of her parentage. It's a sensitive take on the brutality that Bosnian women experienced during the war.

Snow
(2008; drama; directed by Aida Begić)
Immediately after the Bosnian War the women of the Bosniak village of Slavno struggle to survive without their missing menfolk. Then a shocking secret is revealed.

Europe

Playlist

Otkako je Banja Luka postala
Emina Ahmedhodžić-Zečaj
Genre: Sevdah

Voljela sam oči zelene
Hanka Paldum
Genre: Sevdah

Teško meni jadnoj, u Sarajevu samoj
Silvana Armenulić
Genre: Sevdah

Bacila je sve niz rijeku
Indexi
Genre: Rock

Bolje biti pijan nego star
Plavi Orkestar
Genre: Folk/Rock

Nocas je k'o lubenica pun mjesec iznad Bosne
Bijelo Dugme
Genre: Rock

Balkaneros
Goran Bregović
Genre: Truba

Balada o Pišonji i Žugi
Zabranjeno Pušenje
Genre: Rock

Lejla
Hari Mata Hari
Genre: Pop/Sevdah

Love In Rewind
Dino Merlin
Genre: Pop/Folk

Sevdah or *sevdalinka* music is one of the most beloved folk traditions of Bosnia. It's usually slow in tempo and melancholic in tone, like this traditional song about a proud widow from the city of Banja Luka by Emina Ahmedhodžić-Zečaj – one of *sevdah's* most famous proponents.

Bosnia's most famous musician on the international stage, Goran Bregovič was the lead guitarist and creative force behind Yugoslavia's most popular rock band, Bijelo Dugme. His solo career has included acclaimed film soundtracks and folksy numbers such as this knees-up in the raucous brass-dominated Balkan style known as *truba*.

The bridge across the River Drina was completed by Ottoman architect Mimar Sinan in 1577.

Bulgaria

Read List

Pod Igoto
by Ivan Vazov (1889, trans 1894)
In Bulgaria's most famous novel, 'Under The Yoke', Boicho escapes prison and joins the rebellion against Ottoman rule. This powerful epic evokes the fight for Bulgarian identity, in daily life and on the battlefield.

Bai Ganyo
by Aleko Konstantinov (1895, trans 2010)
Exploring modern Bulgaria's difficult birth, this novel's eponymous rose-oil salesman travels around Europe and returns to a country he barely recognises.

East of the West
by Miroslav Penkov (2011)
Nostalgia, loyalty and identity are running themes in Penkov's short stories. This collection offers intriguing glimpses of contemporary Bulgarian life through estranged loves, Orthodox icons and Lenin's corpse.

Sinfonia Bulgarica
by Zdravka Evtimova (2014)
An heiress, a power wife, a masseuse and a struggling waitress lay bare the struggles of womanhood in modern Bulgaria in these candid short stories by multi-award-winning writer Evtimova.

The War of the Letters
by Lyudmila Filipova (2014)
Love, intrigue and religious fervour suffuse this novel set during the First Bulgarian Empire, though the advent and survival of Bulgarian script are its distinctive themes.

Watch List

Lyubimetz 13
(1958; comedy; directed by Vladimir Yanchev)
A man signs a football contract intended for his twin and madcap antics around Sofia ensue – including the brothers' pursuit of the beautiful Elena.

The Goat Horn
(1972; historical drama; directed by Metodi Andonov)
When his wife is murdered by Ottomans, a man trains his daughter to help assassinate the assailants – leaving a horn as the signature of their revenge.

Toplo
(1978; comedy; directed by Vladimir Yanchev)
Residents of an apartment building vow to make conmen contractors finish the job on their central heating. The only problem? The contractors are in jail.

Orkestŭr bez ime
(1982; comedy; directed by Lyudmil Kirkov)
In '*A Band With No Name*', will a group of musicians forsake their political principles for a shot at the big time? Of course not: they caper across Bulgaria instead.

The World is Big and Salvation Lurks Around the Corner
(2008; drama; directed by Stefan Komandarev)
Bikes and backgammon aid a young migrant's recovery from amnesia as he travels back to his home country, Bulgaria, accompanied by his grandfather.

Sofia (this is the Alexander Nevsky Cathedral) is the evocative setting for the film *Lyubimetz 13*.

Playlist

Izlel ye Delyo Haydutin
Valya Balkanska
Genre: Traditional folk

Habibi
Azis
Genre: Chalga (pop-folk)

Middle-Eastern-inflected dance tunes are the soundtrack of modern Bulgaria and Azis is the genre's standout star. Plaintive, love-lorn *Habibi* fills bars across the country, in the inimitable style of this pop royal and nonbinary queer icon.

Koledari
Sofia Boys Choir
Genre: Choral

Nyama bira
Poduene Blues Band
Genre: Blues/Rock

Bate Gojko
Hippodil
Genre: Punk/Ska

Visoko
FSB
Genre: Prog rock

Chekai malko
Upsurt
Genre: Hip-hop

Beautiful Mess
Kristian Kostov
Genre: Pop

Falshiv geroi
Todor Kolev
Genre: Folk/Jazz

Velvety-voiced singer and actor Todor Kolev is one of Bulgaria's most prolific entertainers. This crooning anthem conjures up images of dancing cheek-to-cheek in a Black Sea holiday resort after sundown.

Samo Shampioni
Elitsa Todorova & Stoyan Yankoulov
Genre: Neo-folk

Croatia

Read List

The Return of Philip Latinowicz
by Miroslav Krleža (1932, trans 1959)
In a book that's been described as the first modern Croatian novel, a modernist painter returns to his hometown hoping for inspiration, but finds tangled relationships and stubborn intellectuals.

Cyclops
by Ranko Marinković (1965, trans 2010)
On the eve of WWII, a Zagreb theatre critic decides to starve himself to avoid fighting, and meets actors, sex workers and poets in an intense account of a world on the edge of destruction.

How We Survived Communism and Even Laughed
by Slavenka Drakulić (1991)
Drakulić grew up in Rijeka and Zagreb, and her fascinating study-cum-memoir explores daily life for women and families under communism, from fashion to turnips.

Baba Yaga Laid an Egg
by Dubravka Ugresić (2007, trans 2011)
Baba Yaga is a child-catching witch; in this darkly funny but very human work, Ugresić uses this ancient Slavic legend to explore the lives of disconnected women.

No-Signal Area
by Robert Perišić (2020)
Stange, observant account of a forgotten community: two cousins move to a provincial post-Communist town to reopen a turbine factory and become entangled with the frustrated locals and a criminal gang.

Watch List

Train Without a Timetable
(1959; drama; directed by Veljko Bulajić)
Bulajić is Croatia's most famous director, known for his patriotic Yugoslav war films. His debut is a more nuanced affair, following Dalmatian villagers who are moved to an area of undeveloped farmland after WWII.

H-8
(1958; drama; directed by Nikola Tanhofer)
As a bus and a truck travel between Zagreb and Belgrade in the pouring rain, Tanhofer tells the occupants' stories in this atmospheric film that's often rated as Croatia's best.

Mondo Bobo
(1997; drama; directed by Goran Rušinović)
This stylish black-and-white crime drama was Croatia's first independent feature. The visuals recall Godard, while the tangled plot is inspired by real-life Croatian killer and fugitive Vinko Pintarić.

Marshal Tito's Spirit
(1999; comedy; directed by Vinko Brešan)
Life is quiet on the island of Vis, but when former Yugoslav leader Marshal Tito's ghost appears, tourists and the police soon follow in this likeably eccentric comedy.

The High Sun
(2015; drama; directed by Dalibor Matanić)
Two villages, one Serbian and one Croatian, form the setting for Matanić's beautifully shot study of the scars of war, which follows stories of trauma, love and resentment across 30 years.

Stiniva beach is a highlight of Vis, the island star of the comedy *Marshal Tito's Spirit*.

Croatia's queen of pop is Severina, who broke through in the late 1980s. With her glossy, folk-tinged pop, she remains one of the country's biggest stars. She's represented the country in the European Song Contest, starred in films and plays, and is a gossip-magazine staple.

Klapa is a traditional genre that evolved from church choirs. It's particularly associated with Dalmatia, and songs are often concerned with love, landscapes or the sea. Folk is also popular, and is often based on the *tamburica* (a long-necked lute introduced to the country by the Ottomans) or the *citura* (zither).

Playlist

Uno Momento
Severina ft Ministarke
Genre: Pop

Falling In
Petar Dundov
Genre: Techno

Zaspo Janko
Dunja Knebl
Genre: Folk

Nije Sve Tako Sivo
Hladno Pivo
Genre: Rock

Kad Mi Dođeš Ti
Oliver Dragojević
Genre: Pop

Libar
Gibonni
Genre: Pop

Zora Bila
Klapa Intrade
Genre: Klapa

Katarina
Gustafi
Genre: Folk rock

Alles Gut
TBF
Genre: Hip-hop

Goli I Bosi
Elemental
Genre: Hip-hop

Cyprus

Read List

Echoes from the Dead Zone: Across the Cyprus Divide
by Yiannis Papadakis (2005)
In this humane, clear-eyed work of reportage, Papadakis journeys across Cyprus to expose and explore the toll of long-held myths and prejudices in island communities.

Ledra Street
by Nora Nadjarian (2006)
The day-to-day absurdities of life in Nicosia, Europe's last divided capital, are brought to the fore in this first short story collection by Armenian-Cypriot Nadjarian.

Gregory and Other Stories
by Panos Ioannides (2009, trans 2014)
This collection brings together the award-winning works of Greek-Cypriot Ioannides. The stories traverse the sweep of Cypriot history but are bound together by themes of belonging, loyalty and war.

Nicosia Beyond Barriers: Voices from a Divided City
edited by Alev Adil, Aydin Mehmet Ali, Bahriye Kemal & Maria Petrides (2019)
Nicosia's split personality, with the demarcation zone slashing through its centre, is explored by Greek-Cypriot and Turkish-Cypriot writers in this short story collection.

Bitter Lemons of Cyprus
by Lawrence Durrell (1957)
Durrell's memoir of Cypriot life in the mid-1950s remains a highly readable personal account of the end years of British rule and the armed Greek-Cypriot campaign for Enosis, which set the scene for intercommunal violence.

Watch List

Parallel Trips
(2004; documentary; directed by Panikkos Hrysanthou & Derviş Zaim)
A Turkish-Cypriot director and a Greek-Cypriot director journey to each other's side of the Green Line to document the legacy of Cyprus' division.

Akamas
(2006; drama; directed by Panikkos Hrysanthou)
This love story between a Turkish-Cypriot and Greek-Cypriot, set in the turbulent years of intercommunal violence, remains a highly controversial film in Cyprus.

Shadows and Faces
(2010; drama; directed by Derviş Zaim)
Set in the island's violent 1960s, *Shadows and Faces* focuses on how one village is drawn into the conflict; told from the Turkish-Cypriot perspective.

The Story of the Green Line
(2016; drama; directed by Panikkos Hrysanthou)
This story, focused on two soldiers who face each other across the Green Line, explores themes of displacement and passed-down generational revenge.

Smuggling Hendrix
(2018; comedy-drama; directed by Marios Piperides)
The absurdities of partition are highlighted in this film about a dog that runs across the Green Line and owner Yiannis's increasingly ludicrous attempts to bring him back home.

Europe

Playlist

Kozan Marşı
Mehmet Ali Sanlıkol & DÜNYA ensemble
Genre: Folk

The System
Monsieur Doumani
Genre: Modern folk

Piano Sonata III: Folk Dance in Cypriot Style
Christos Tsitsaros
Genre: Classical

Bana da Söyle
Ziynet Sali
Genre: Turkish pop

Rotten Luck
Trio Tekke
Genre: Modern rebetiko

Dolama
Grup Net
Genre: Folk

Eleni Elenara Mou
Michalis Terlikkas
Genre: Folk

Ta Ryalia
Michalis Violaris
Genre: Modern laika

Rain
Vasiliki Anastasiou
Genre: Modern folk

I-Huzum
Tri-Cool-Ore
Genre: Jazz fusion

Critically acclaimed Nicosia three-piece Monsieur Doumani have released three albums since forming in 2012. They're known for their distinctive modern-folk style, using traditional Cypriot instruments, and songwriting that often focuses on problems in Cypriot society.

One of the island's most successful musical exports, Michalis Violaris has been releasing music since the 1960s when he moved to Greece. His 1973 song *Ta Ryalia* is known for being the first hit song in Greece, sung in the Cypriot-Greek dialect.

Lines of division in Nicosia are a significant theme in many books and movies from Cyprus.

Czech Republic

Read List

The Good Soldier Švejk
by Jaroslav Hašek (1921-1923, trans 1997)
Hašek's hilarious account of the wanderings of a hapless Czech soldier during WWI in Austria-Hungary could easily top any list of Czech literary masterpieces.

Description of a Struggle
by Franz Kafka (1936, trans 1958)
Kafka's work rarely references specific place names but this long short story is full of recognisable Prague locations; it convinced Kafka's friend and editor Max Brod that he should pursue a career in writing.

I Served the King of England
by Bohumil Hrabal (1971, trans 1989)
A classic caper following the ups and downs of an up-and-coming hotelier in Nazi-occupied and then early Communist Czechoslovakia.

The Unbearable Lightness of Being
by Milan Kundera (1984)
Exiled Kundera wrote this fine novelistic outpouring in French and it was published in English before appearing in Czech. It charts intellectual Prague society in 1968 between the Prague Spring and Soviet invasion.

Love and Garbage
by Ivan Klíma (1986, trans 2002)
An artist is forced to become a refuse collector in Communist Prague in this hugely successful book that's both a thought-provoking existential meditation and illustration of the challenges of life in the regime.

Watch List

Closely Observed Trains
(1966; comedy-drama; directed by Jiří Menzel)
This Academy Award foreign-language film winner features Miloš, a newly-employed guard at a railway station in Nazi-occupied Czechoslovakia during WWII: Czech New Wave at its best.

The Fireman's Ball
(1967; comedy; directed by Miloš Forman)
Forman's last movie before leaving Czechoslovakia for the USA is a shrewd, tongue-in-cheek look at Communist party ineptitude, set around preparations for a fire department's yearly ball in a small Czech town.

My Sweet Little Village
(1985; comedy; directed by Jiří Menzel)
This loveable, laugh-rich account of Czech small-town goings-on revolves around a truck-driving duo: sweet-natured simpleton Otik and his sidekick Pávek.

The Elementary School
(1991; comedy-drama; directed by Jan Svěrák)
Post-WWII Prague: Eda, a schoolboy with an active imagination, is in a terribly behaved class that teacher Igor is drafted in to discipline. Only Igor is not what he initially seems...

Cosy Dens
(1999; comedy-drama; directed by Jan Hřebejk)
A coming-of-age comedy tinged with the drama of 1968's Prague Spring. With wit and originality it unveils contrasting attitudes to this event.

Prague's distinctive character appears in books by Franz Kafka and Milan Kundera.

Gott, Czech music's most internationally renowned figure, dons 19th-century military garb in this 1960s classic. Never was singing about loading a cannon so feel-good, and this song shows how the man's universal appeal could transcend every decade up to the 2010s.

Toxika is a trance-like track imbibed with haunting violin from PPU's first studio album, *Egon Bondy's Happy Hearts Club Banned*. PPU were a properly underground group and regularly persecuted by Czech authorities. Had they not been, they would undoubtedly have achieved infinitely more acclaim.

Playlist

Europe

Bum Bum Bum
Karel Gott
Genre: Classic pop

Opus 72 No 2 in E Minor from the Slavonic Dances
Antonín Dvořák
Genre: Classical

The Bugatti Step
Jaroslav Ježek
Genre: Jazz

Bratříčku, zavírej vrátka
Karel Kryl
Genre: Folk

Trouba (Release Me)
Lucie Bílá
Genre: Pop

Nonstop
Michal David
Genre: Electro-pop

Toxika
Plastic People of the Universe (PPU)
Genre: Prog rock

Andělé
Wanastowi Vjecy
Genre: Rock

Missing Feeling Nothing
Please The Trees
Genre: Indie/Rock/Folk

Mezi horami
Čechomor
Genre: Folk rock

Denmark

Read List

The Little Mermaid
by Hans Christian Andersen (1837)
Andersen was a prolific storyteller, unafraid to eschew the happy ending. In this iconic fairy tale (one of 156 he penned), a mermaid yearns for a human soul.

Mercy
by Jussi Adler-Olsen (2007, trans 2011)
This classic of the Nordic Noir genre sets the stage for the Department Q series, in which Copenhagen's coldest cases are investigated by a flawed detective. It was published as *The Keeper of Lost Causes* in the US.

Noma: Time and Place in Nordic Cuisine
by René Redzepi (2010)
New Nordic cuisine swept the culinary world in the 2010s and this cookbook from Redzepi, the movement's poster child, beautifully illustrates the locavore commandments. The brave can seek ingredients and attempt its recipes.

Miss Smilla's Feeling for Snow
by Peter Høeg (1992, trans 1993)
On the surface, this novel is a suspenseful mystery partly set in the frigid north (and later made into a film); below the surface, it's an exploration of the complex post-colonial relationship between Denmark and Greenland.

We, the Drowned
by Carsten Jensen (2006, trans 2011)
Danish seafaring history is at the fore in this titanic tale, which follows four generations of salty sailors from the port town of Marstal.

Watch List

Another Round
(2020; drama; directed by Thomas Vinterberg)
Would everyone benefit from more alcohol in their bloodstream? Four high-school teachers test a wild theory in this reflection on friendship, ageing and the Danish drinking culture.

Babette's Feast
(1987; drama; directed by Gabriel Axel)
An Oscar-winning tale that has a single, splendid meal at its heart, and a feel-good message of food as a language of love that transcends cultural and religious divides.

Klown
(2010; comedy; directed by Mikkel Nørgaard)
Crude, lewd and cheerfully offensive – plus with some surprising heart – *Klown* offers revealing insight into Danish humour. It's based on a long-running TV series.

In a Better World
(2010; drama; directed by Susanne Bier)
Winner of Best Foreign Language Film at the Oscars, the plot of this drama veers between school bullying, infidelity, bereavement, revenge and warlords, questioning cosy stereotypes about Denmark.

Festen
(1998; comedy drama; directed by Thomas Vinterberg)
Centred on a family gathering to celebrate a milestone birthday, this film – also known as '*The Celebration*' blends farce and tragedy with precision. It ushered in a new maverick style of film-making known as Dogme 95.

Breezy Copenhagen, such as Nyhavn here, has inspired everything from cookbooks to whodunits.

Fly on the Wings of Love won the Eurovision Song Contest for Denmark in 2000. The song wasn't expected to score highly, as it was an old-fashioned ballad performed by two of the oldest performers to have entered the contest.

Among the catchy Phlake musical fusion of R&B, pop, hip-hop and funk, listen out for some unexpectedly offbeat lyrics in songs like *Pregnant* and *Ikea Episodes*.

Europe

Playlist

Final Song
MØ
Genre: Pop

Fly on the Wings of Love
Olsen Brothers
Genre: Easy listening

Øde Ø
Rasmus Seebach
Genre: Pop

Barbie Girl
Aqua
Genre: Pop

Fascination
Alphabeat
Genre: Pop

Let Me Think About It
Ida Corr
Genre: Dance

Lækker
Nik & Jay
Genre: Hip-hop

Angel Zoo
Phlake
Genre: R&B

Drunk in the Morning
Lukas Graham
Genre: Pop

Hollow Talk
Choir of Young Believers
Genre: Chamber pop

England

Read List

Hamlet
by William Shakespeare (1603)
Perhaps Shakespeare's greatest play, this classic of murder, revenge and princely procrastination is packed with immortal soliloquies.

Emma
by Jane Austen (1815)
Austen's elegant, playful novel follows a match-making heroine who's not quite as clever as she thinks she is. It's full of wit, romance and the beautifully evoked lives of the country gentry.

Great Expectations
by Charles Dickens (1861)
The characters in Dickens' melodramas feel as vivid today as they did in his Victorian heyday. Memorable, unhinged Miss Havisham helps make this the pick of the bunch.

Wolf Hall
by Hilary Mantel (2009)
Mantel brings Henry VIII's adviser Thomas Cromwell to life in a complex but deeply readable epic that plunges the reader into a Renaissance world of faith, power and intrigue. It's the first in a superb trilogy.

Girl, Woman, Other
by Bernardine Evaristo (2019)
Twelve lives intertwine in a funny, confident and compassionate book that touches on Twitter spats, marriage, theatre, youth and betrayal. It's about race, sex and politics – and how we relate to each other.

Watch List

Kind Hearts and Coronets
(1949; comedy; directed by Robert Hamer)
The best of the Ealing Comedies of the 1940s and '50s, this is a charming, understated account of serial murder, with Dennis Price popping off members of his aristocratic family – all played by Alec Guinness – one by one.

Lawrence of Arabia
(1962; drama; directed by David Lean)
Lean's biopic of troubled British officer T E Lawrence (played by Peter O'Toole) is the ultimate epic, a stunningly shot tale set among the sands of WWI Syria and Jordan.

Monty Python's Life of Brian
(1979; comedy; directed by Terry Jones)
Brian is not the messiah – he's a very naughty boy – in this uproarious, endlessly quotable riff on religion from the legendary comedy troupe.

This is England
(2006; drama; directed by Shane Meadows)
Teenage Shaun starts hanging with a friendly gang of skinheads, before racism rears its shaved head. Meadows's dark but charismatic film is a bittersweet look at social change and 1980s youth culture.

Skyfall
(2012; drama; directed by Sam Mendes)
James Bond has been battling henchmen, wooing women and drinking martinis on the payroll of the British Secret Intelligence for 70 years, and this stylish Daniel Craig outing – with a Highlands-set finale – is one of the best.

Europe

English music has often looked to the US, both for inspiration and record sales. The Beatles, the Stones and the Kinks led a 'British Invasion' in the 1960s; in the 1980s, the US charts were raided again, by the Human League, Dire Straits and Bananarama.

From blues-rock to hip-hop, England has a habit of taking genres, giving them a twist and selling them back to the world. So it was with house and techno, which were born in the US but gained huge exposure through Britain's 1980s and '90s club scene.

Every Janeite should make a pilgrimage to Chawton in Hampshire, where Jane Austen lived in this cottage.

Playlist

Anarchy in the UK
The Sex Pistols
Genre: Punk

Gimme Shelter
The Rolling Stones
Genre: Rock

Unfinished Sympathy
Massive Attack
Genre: Trip-hop

Land of Hope and Glory
Edward Elgar
Genre: Classical

Shape of You
Ed Sheeran
Genre: Pop

Heroes
David Bowie
Genre: Rock

Voodoo Ray
A Guy Called Gerald
Genre: House

I Do Like To Be Beside the Seaside
Mark Sheridan
Genre: Music hall

Rehab
Amy Winehouse
Genre: Soul

Kate Bush
The Hounds of Love
Genre: Rock

IN DEPTH

The Beatles

from ENGLAND

Left to right: Abbey Road's zebra crossing; The Beatles cut their teeth in Liverpool's Cavern Club; the band's last live gig on the roof of the Apple building in 1969.

In ten years, the Beatles changed the world. From 1960 to 1970, four young men from Liverpool recorded 17 British and 20 US number one singles. They spearheaded the 'British Invasion' of the US, played the first ever stadium rock gig (at New York's Shea Stadium) and became flag-bearers for '60s counterculture. Oh, and they're the biggest-selling band ever, with 600 million album sales under their belt. Not bad for a group who were famously rejected by record company Decca with the words, 'guitar groups are on the way out'.

The Beatles were an incredible pop band, marshalling teen delirium across the world in their suits and mop-top haircuts. But their accessible look masked a fearsome live show – honed during gruelling residencies playing R 'n' B covers in Liverpool and Hamburg. And their ear for a melody was married to a creative impulse that was genuinely revolutionary. The prolific partnership between John Lennon and Paul McCartney established the idea that bands should write their own songs. Rubber Soul set the blueprint for psychedelia and 'Helter Skelter' arguably invented heavy metal. As the band inched towards its acrimonious end in the late '60s, they still making some of the most hummable pop songs ever, while using the studio as an instrument, reversing recordings and pioneering the use of synthesizers. No one has matched them since.

Estonia

Read List

The Czar's Madman
by Jaan Kross (1978, trans 1992)
Estonia's best-known writer channelled his years of Soviet imprisonment to create this psychological tour de force about a 19th-century nobleman with an oppressive ruler.

Vargamäe
by A H Tammsaare (1926, trans 2019)
Volume one of Tammsaare's renowned pentalogy *Truth & Justice* grapples with societal upheavals, the decline of morality and other big topics while charting Estonian village life from the late 1800s to the 1905 revolution.

Selected Poems
by Jaan Kaplinski (2011)
Forests and nature are central to Estonian identity and feature prominently in Kaplinski's elegant, Zen-like poetry. His wide-ranging musings have undercurrents of Buddhism, Eastern philosophy and Celtic mythology.

Kalevipoeg
by Friedrich Reinhold Kreutzwald (1853)
Giants, sorcerers and talking hedgehogs star in this magnificent, *Odyssey*-like epic poem. It's a national classic, drawing on Estonia's ancient oral folk traditions.

Estonia and the Estonians
by Toivo Raun (2nd edition, 2002)
A concise but well-written overview of Estonian history from Paleolithic times through independence following the USSR's collapse, with special focus on the (heartbreaking) WWII era and Soviet era.

Watch List

Truth and Justice
(2019; drama; directed by Tanel Toom)
A blockbuster in Estonia, this ambitious, pan-generational 19th-century saga depicts a tenacious farm family working the land amid formidable social and environmental challenges.

Spring
(1969; drama; directed by Arvo Kruusement)
Rated by Estonians as one of the nation's best films, *Spring* revolves around a group of schoolboys – their friendships, betrayals and first loves portrayed in a bittersweet coming-of-age tale.

The Singing Revolution
(2006; documentary; directed by James Tusty & Maureen Castle Tusty)
The deeply moving story of Estonians gathering by the hundreds of thousands to sing together in hopes of gaining their freedom from the Soviet Union.

Tangerines
(2013; drama; directed by Zaza Urushadze)
A simple but powerful Estonian-Georgian film about two enemy soldiers who are thrown together and learn to recognise one another's humanity and the futility of war.

Names Engraved in Marble
(2002; drama; directed by Elmo Nüganen)
Against the backdrop of Estonia's war for independence (1918–1920), schoolboys join the army and face the gravest sacrifice: risking their lives for their homeland.

Find KGB spy equipment at the gripping Hotel Viru and KGB Museum in Tallinn.

One of the world's greatest living composers, Arvo Pärt is famed for an austere, minimalist music style known as *tintinnabuli*. These atmospheric, meditative compositions were influenced by Gregorian chant and Renaissance polyphony, and sound at once both ancient and modern.

Estonia's folk music has deep roots (hymn-like runic songs date back to the 12th century), which live on in countless modern interpreters. Inspired by nature and rural life, Mari Kalkun sings dreamy lyrics written by local poets and plays the harp-like *kannel*.

Playlist

Pulsar, Part 1
Estrada Orchestra
Genre: Jazz

Spiegel im Spiegel
Arvo Pärt
Genre: Classical

(In the End) There's Only Love
Ewert and the Two Dragons
Genre: Indie-rock

Savages
Kerli
Genre: Goth-pop

Vaid Vaprust
Metsatöll
Genre: Folk-metal

Hingede Öö
Maria Minerva
Genre: Nu-disco

Mõstsavele Mäng
Mari Kalkun
Genre: Folk

Honeymooning
Holy Motors
Genre: Indie-rock

Nelgid
Vaiko Eplik
Genre: Pop

Kuukene
Trad.Attack!
Genre: Folk

Finland

Read List

Kalevala
by Elias Lönnrot (1835, trans 1888)
Finland's national epic is a full-on Scandi folk feast, with deep, dark forests, giants and mythical beasties aplenty. It is based on the legends, oral runes and poems Elias Lönnrot collected while travelling through eastern Finland.

Seven Brothers
by Aleksis Kivi (1870, trans 1929)
One of Finland's most feted literary treasures, this down-to-earth tale of hapless brothers, who inherit the family farm and escape conventional life in the forest, embodies the birth of Finnish national consciousness.

Let the Northern Lights Erase Your Name
by Vendela Vida (2007)
When a young woman unearths a family secret, she leaves New York to journey to Lapland in search of her Sámi origins and discovers her true self.

The Year of the Hare
by Arto Paasilinna (1975, trans 1995)
A bizarre but gripping tale about a jaded, frustrated, city-dwelling journalist, who escapes into the wilderness after his car hits a hare on a country road.

The Summer Book
by Tove Jansson (1972, trans 2003)
Finland's beloved Moomin creator deviates into adult fiction in this magical, life-affirming novel set on a tiny Finnish island, chronicling the relationship between an elderly artist and her six-year-old granddaughter.

Watch List

The White Reindeer
(1952; horror; directed by Erik Blomberg)
A horror fairy tale shot in the snowy fells of Lapland, about a newlywed woman who seeks a shaman's help and gets turned into a vampiric white reindeer.

Man Without a Past
(2002; comedy drama; directed by Aki Kaurismäki)
The second part of Kaurismäki's darkly comic Finland trilogy involves a man in Helsinki who loses his memory after being mugged and moves into a shipping container.

Moomins on the Riviera
(2014; animated comedy; directed by Xavier Picard & Hanna Hemilä)
Based on comic strips by Tove and Lars Jansson, here the much-loved Moomins find a wrecked pirate ship in the fjord and use the booty to holiday in the French Riviera.

Steam of Life (Miesten vuoro)
(2010; documentary-drama; directed by Joonas Berghäll & Mika Hotakainen)
This riveting documentary turns up the heat with its frank, comic and culturally insightful tour of the country's saunas: a fascinating look at Finland's sauna obsession.

The Unknown Soldier
(2017; war-drama; directed by Aku Louhimies)
This award-winning movie is the third adaptation of the Finnish novel by Väinö Linna (considered a national legacy), recounting the story of an infantry unit in the Finnish Continuation War against the Soviet Union.

Finland's Aurora Borealis swirls across its northern skies above the lands of the Sámi people.

Playlist

Levottomat Jalat
Hassisen Kone
Genre: Rock

Finlandia, Op 26
Jean Sibelius
Genre: Classical

Written for the Press Celebrations of 1899, a covert protest against increasing censorship from the Russian Empire, *Finlandia, Op 26* is a moody, powerful symphonic poem and the alternative national anthem. It's Finland put to music.

Gidda Beaivvas
Angelit
Genre: Folk

Katjusha
Leningrad Cowboys
Genre: Pop/Rock

Buried Alive by Love
HIM
Genre: Finnish Gothic rock

Bittersweet
Apocalyptica
Genre: Symphonic metal

Bájan Riegáda
Wimme Saari
Genre: Finnish Sámi yoik

Wimme Saari is one of the best-known Finnish Sámi artists, combining *yoik* (a chant sung a cappella, which evokes a person or place with spiritual importance in Sámi culture) with modern improvisations, usually to a techno-ambient accompaniment.

Seelinnikoi
Värttinä
Genre: Folk

Chasing Highs
Alma
Genre: Pop

Ramblin
Jukka Tolonen
Genre: Jazz

France

Read List

The Devil's Pool
by George Sand (1846)
The pen name of Amantine Lucile Aurore Dupin, Sand was, in her time, as popular and revered as contemporary Victor Hugo. *The Devil's Pool* is the first in a quartet of pastoral novels about her childhood in central France.

Remembrance of Things Past
by Marcel Proust (1913-1927, trans 1922)
Released in stages, this leviathan novel examines the narrator's aristocratic life in turn-of-the-century France. It's often cited as the greatest novel of all time.

Wind, Sand and Stars
by Antoine de Saint Exupéry (1939)
This memoir is a pilot's love letter to flight – as well as an account of survival following a crash in the Egyptian desert, and a treatise on the true meaning of life. The author was shot down near Marseille in WWII.

The Stranger
by Albert Camus (1942, trans 1946)
The story of a shipping clerk in French Algeria and his life in the aftermath of his mother's death. It's a novel that has continued to arouse strong feelings into the 21st century.

Serotonin
by Michel Houellebecq (2019)
This novel by France's most controversial contemporary writer explores a depressed bureaucrat's personal life in Paris. The depiction of armed blockades on Normandy roads foreshadow the *gilet jaune* (yellow vests) movement.

Watch List

Monsieur Hulot's Holiday
(1953; comedy; directed by Jacques Tati)
Set among the resorts of the Atlantic Coast, and rich in visual gags, this warm-hearted comedy stars director Tati as Monsieur Hulot – a member of a new post-war holidaymaking class.

The Umbrellas of Cherbourg
(1964; musical; directed by Jacques Demy)
All dialogue is sung – like an opera – in this celebrated musical. A 17-year-old girl working in a Cherbourg umbrella boutique has her life turned upside down when her boyfriend departs to serve in the Algerian War.

Three Colours: Blue
(1993; drama; directed by Krzysztof Kieślowski)
The first in a trilogy where each film explores one of the French revolutionary ideals (*liberté*, *égalité* and *fraternité*) – *Blue* stars Juliette Binoche as a Parisian woman coming to terms with the death of her family in a car accident.

La Haine
(1995; drama; directed by Mathieu Kassovitz)
A visceral and often hilarious account of 19 hours in one of the poorer *banlieues* of Paris, *La Haine* sees three friends plotting vengeance on police after a friend is attacked.

Hidden
(2005; thriller; directed by Michael Haneke)
A complex, intense thriller, *Hidden* stars Daniel Auteil and Juliette Binoche as a Parisian couple who discover their life is under constant surveillance.

Jacques Tati scouted locations from Dunkirk to Biarritz (above) for his movie *Monsieur Hulot's Holiday*.

Sometimes unfairly viewed as a novelty artist outside his native France, at home Gainsbourg remains a deeply revered figure for the diversity of his musical output, as well as his sharp, multilayered lyrics.

The first French rapper to achieve international fame, Senegal-born MC Solaar offers an immigrant's perspective on modern France and its social divides. *La Belle Et Le Bad Boy* is an iconic song telling the story of two starstruck lovers, forced apart by violent crime.

Europe

Playlist

Je suis venu te dire que je m'en vais
Serge Gainsbourg
Genre: Pop

Tilted
Christine and the Queens
Genre: Pop

Genesis
Justice
Genre: Electronic

La Vie En Rose
Édith Piaf
Genre: Cabaret

Les Champs-Élysées
Joe Dassin
Genre: Pop

L'Aventurier
Indochine
Genre: Rock

La Belle Et Le Bad Boy
MC Solaar
Genre: Rap

Cendrillon
Téléphone
Genre: Rock

Petit Frère
Iam
Genre: Rap

La Bohème
Charles Aznavour
Genre: Singer-songwriter

IN DEPTH

New Wave Cinema

from FRANCE

©BERNARD ALLEMANE \ INA VIA GETTY IMAGES

Left to right: François Truffaut on the set of *Stolen Kisses*; director Agnès Varda as president of the Cannes Film Festival jury; Jean-Luc Godard with Anna Karina and Jean-Paul Belmondo in 1965.

© PASCAL LE SEGRETAIN / GETTY IMAGES

The French New Wave sent shockwaves around the world in the late 1950s and '60s – and its legacy is still visible in international cinema today. It emerged from a much-changed, post-WWII country searching for a new cinematic language – taking inspiration from other 1950s movements, such as the everyday stories of Italian neorealism, and the grittiness of American noir. At the New Wave's own genesis was the Parisian magazine *Cahiers du Cinéma*, in which critics emphasised the importance of the auteur – of directors such as Agnès Varda, François Truffaut and Jean-Luc Godard as artists, as something to be valued above plot, acting or filmmaking craft.

Bold styles arose: realistic dialogue, handheld camerawork and jump cuts – while the existential angst of modern French life was an enduring theme. Some of the greatest examples include François Truffaut's *Les Quatre Cent Coups* (*The 400 Blows*) – a coming-of-age story about a misunderstood boy growing up in post-war Paris, based on the director's own experiences – and Jean Luc Goddard's *À bout de souffle* (*Breathless*) – a pacy thriller that begins with its lead a hot-wiring a car in Marseille, and murdering a policeman on a country road. The New Wave has been cited as an influence on Tarantino and Scorsese – its revolutionary approach is unmistakably French.

Georgia

Read List

The Knight in the Panther's Skin
by Shota Rustaveli (circa 1200)
This epic poem is Georgia's most sacred work of literature – citizens once learned its verses by heart. It tells the story of heroes Avtandil and Tariel and their chivalrous deeds in faraway India and Arabia.

***Host and Guest*,**
by Vazha-Pshavela (1893)
A poem that still speaks to intra-religious tensions in the Caucasus today, *Host and Guest* is the story of two men whose fates unravel when the Christian guest is invited to stay with his host's Muslim community.

The Right Hand of the Grand Master
by Konstantine Gamsakhurdia (1939)
A Georgian legend told within the framework of a western European novel, Gamsakhurdia's tale concerns the building of the Svetitskhoveli Cathedral, just outside Tbilisi, against a backdrop of feuding mountain tribes.

Granny, Iliko, Illarion, and I
by Nodar Dumbadze (1960, trans 1985)
A semi-autobiographical novel set in rural Georgia in the years around WWII, where the menfolk have departed for the battlefront.

Data Tutashkhia
by Chabua Amirejibi (1975, trans 1985)
A wildly popular novel during the Soviet era, chronicling the escapades of the title character – a Robin Hood figure in the mountains of Tsarist-era Georgia.

Watch List

Father of a Soldier
(1964; war-drama; directed by Revaz Chkheidze)
Among the most celebrated Soviet war movies, *Father of a Solider* stars Sergo Zakariadze as an elderly wine farmer who decides to join the army, and ends up fighting beside his son on the battlefields of Germany.

The Plea
(1967; art house; directed by Tengiz Abuladze)
A cinematic rendering of the poems of Vazha-Pshavela, *The Plea* portrays a symbolic battle between good and evil, with striking chiaroscuro visuals.

Tangerines
(2013; drama; directed by Zaza Urushadze)
Set during the War in Abkhazia (1992-1993) – a breakaway region of Georgia – this film sees an Estonian tangerine farmer take in wounded soldiers from both sides of the conflict.

My Happy Family
(2017; drama; directed by Nana Ekvtimishvili)
Fiftysomething teacher Manana resolves to leave her husband, children and home to start a new life, in a movie that delves deep into Georgian family values.

And Then We Danced
(2019; drama; directed by Levan Akin)
Born in Sweden to Georgian parents, Akin received worldwide acclaim for this story of two male dancers in the National Georgian Ensemble, whose love affair must be kept secret in their ultraconservative homeland.

Europe

Playlist

Miniatures for Violin & Piano
Giya Kancheli
Genre: Modern classical

Raise Your Head & Smell the Air
Nikakoi
Genre: Electronica

Once in the Street
Nino Katamadze
Genre: Jazz

Sunrise
Okinawa Lifestyle
Genre: Electronica

Beware the Silence
Psychonaut 4
Genre: Black metal

Isev Da Isev
Mgzavrebi
Genre: Folk-pop

Tsintskaro
Hamlet Gonashvili
Genre: Folk/Choral

Me Da Shen
Megi Gogitidze
Genre: Pop

Mravaljamieri
Ensemble Shavnabada
Genre: Folk/Choral

Chakrulo
Rustavi Choir
Genre: Folk/Choral

Despite being resident in Belgium for much of his later life, the recently deceased Giya Kancheli was Georgia's most lauded classical composer of modern times, integrating elements of Georgian folk music into his compositions.

The Right Hand of the Grand Master revolves around Svetitskhoveli Cathedral in historic town of Mtskheta.

The Georgian folk song that has travelled most widely, the haunting *Tsintskaro*, describes an encounter with a beautiful girl by a spring. It has featured on film soundtracks, and also in the closing moments of Kate Bush's album, *Hounds of Love*.

Germany

Read List

The Tin Drum
by Günter Grass (1959, trans 1961)
The Tin Drum is Nobel Prize–winning author Grass' highly acclaimed first novel. It recounts Germany's 20th-century history told from the perspective of a child, Oskar, who refuses to grow up.

Austerlitz
by WG Sebald (2001)
Austerlitz tells the powerful and haunting story of a man's search for his identity after being sent to England in 1939 on the Kindertransport as a five-year-old boy.

They Divided the Sky
by Christa Wolf (1963, trans 1965)
Wolf weaves a heartbreaking love story set in the early 1960s of a young couple's relationship struggling to survive amid the construction of the Berlin Wall.

Berlin Alexanderplatz
by Alfred Döblin (1929, trans 1931)
Set in 1920s Berlin, this novel is widely regarded as a literary masterpiece of the Weimar Republic. It tells the story of petty criminal Franz Biberkopf attempting to set himself straight after a stint in prison.

Origin (Herkunft in German)
by Saša Stanišić (2019)
In this autobiographical 2019 German Book Prize winner, Bosnian-German writer Stanišić explores complex themes of roots and identity through his memories as his grandmother loses her own memories to dementia.

Watch List

The Lives of Others
(2006; drama/thriller; directed by Florian Henckel von Donnersmarck)
An honest and insightful Oscar-winning film portraying the sinister operations of the East German secret police, the Stasi, and its impact on ordinary citizens.

Good Bye, Lenin!
(2003; tragicomedy; directed by Wolfgang Becker)
A witty story of a son recreating the GDR to save his sick mother from the shock of learning about the fall of the Berlin Wall.

Metropolis
(1927; sci-fi; directed by Fritz Lang)
Lang's silent feature visual masterpiece was a groundbreaking and hugely influential film set in a towering, futuristic dystopia in the year 2000.

Downfall
(2004; drama; directed by Oliver Hirschbiegel)
Nominated for an Oscar for Best Foreign Film, *Downfall* attracted controversy for its portrayal of the human side of Hitler in his final days in the bunker beneath Berlin.

Das Boot
(1981; action/war; directed by Wolfgang Petersen)
Nominated for several Oscars, '*The Boat*' is heralded as one of the great war movies of all time. It takes viewers into the incredibly tense story of a German U-boat and its crew during WWII.

Europe

Playlist

99 Luftballons
Nena
Genre: Pop

Lili Marlene
Marlene Dietrich
Genre: Jazz/Pop

Originally written as a poem by a German soldier during WWI, the love song *Lili Marleen* was sung by German American Hollywood songstress Marlene Dietrich and became something of an anthem of remembrance. It was popular with both sides of the conflict during WWII.

A Midsummer Night's Dream
Felix Mendelssohn
Genre: Classical

Links 2,3,4
Rammstein
Genre: Heavy metal

Kaltes Klares Wasser
Malaria!
Genre: New wave/Post-punk

German Requiem
Johannes Brahms
Genre: Classical

Major Tom
Peter Schilling
Genre: Synth-pop

Autobahn
Kraftwerk
Genre: Electronic

The hit title track from Kraftwerk's 1975 album was the band's first song to have lyrics (sung in German) and made it to #25 on the US Billboard Hot 100 chart.

Fremd im Eigenen Land
Advanced Chemistry
Genre: Hip-hop

Tage Wie Diese
Die Toten Hosen
Genre: Rock

The Berlin Wall, here showing what was once the East German side, is pivotal to *The Lives of Others*.

IN DEPTH

Love Parade

from GERMANY

The epicentre of the world's electronic/techno/ dance music scene each year was Berlin's iconic Love Parade music festival – until tragedy struck in 2010, ending the event's two-decade run. Created in 1989 by DJ Motte (Matthias Roeingh) and his girlfriend Danielle de Picciotto, it was initially intended as a political demonstration spreading peace and love with the motto 'Friede, Freude, Eierkuchen' ('Peace, Joy, Pancakes') and was originally held on Berlin's main shopping strip, Kurfürstendamm. The first parade included one float followed by a car blaring techno music, and from there its popularity steadily grew and it went on to

become one of the largest music festivals in Europe and an annual pilgrimage for every electronic music lover and techno rave fan.

The festival will never be held again after it came to an end following a horrific deadly stampede inside a congested tunnel at the 2010 Love Parade festival in the city of Duisburg. Twenty-one people were killed and hundreds more were injured in the crush. Festival organisers were on trial over criminal negligence allegations before the court closed the investigation in 2020 with no convictions.

Left to right: techno producer DJ Motte, the founder of the Love Parade; a raving reveller; dancers swirl around Berlin's Victory Column in 1999.

Greece

Read List

The Odyssey
by Homer (8th century BCE)
In a battle between *The Odyssey* and *The Iliad*, this wins for having more monsters and more narrative drive. It's an epic poem about Greek hero Odysseus's long trip home, and one of the cornerstones of Western literature.

The Dialogues of Plato
by Plato (4th century BCE)
Various editions collect Plato's dialogues, letting you start with the more approachable early works before diving into *The Republic's* exploration of justice and society.

The Late-Night News
by Petros Markaris (1995, trans 2004)
A troubled detective investigates a murdered Albanian couple and follows a trail through media corruption and sex trafficking. Markaris specialises in page-turning thrillers fuelled by rage at Greece's inequality.

Rebetiko
by David Prudhomme (2012)
This wild and playful graphic novel is a great introduction to *rebetiko*, Greece's urban folk/blues music hybrid of the 1930s that enraged the authorities, as well as a fine tale of nightlife and companionship.

Austerity Measures: The New Greek Poetry
edited by Karen Van Dyck (2016)
Greek verse for the modern age: austerity, identity, language, migration and consumerism jostle in this collection that celebrates Greece's poetry renaissance.

Watch List

Stella
(1955; drama; directed by Michael Cacoyannis)
When a folk singer meets a football player, her independent life is threatened. This feminist melodrama made Melina Mercouri a star – and helped usher in Greek cinema's golden age.

The Ogre of Athens
(1956; drama; directed by Nikos Koundouros)
A bank clerk is mistakenly identified as a famous criminal and decides to live the lie in a film noir that flopped at the time, but is now considered one of Greece's finest.

Zorba the Greek
(1964; comedy drama; directed by Michael Cacoyannis)
This multiple Oscar winner and commercial smash hit stars Anthony Quinn as a Greek peasant, Alan Bates as an uptight Englishman, and a riotous *bouzouki*-toting soundtrack.

Landscape in the Mist
(1988; drama; directed by Theo Angelopoulos)
Two children travel through Greece to Germany to find a father who may not even exist, in a dark, tender and beautifully shot odyssey.

Dogtooth
(2009; drama; directed by Yorgos Lanthimos)
Lanthimos's compelling and deeply strange film tells of a family raised in isolation by their controlling father, complete with false words and phantom relatives.

Europe

Athens is the setting for many a Greek drama, classical or contemporary, including *The Ogre of Athens* film.

One of the great soundtrack composers and a pioneer of electronic music, Vangelis performed in progressive rock band Aphrodite's Child (with Demis Roussos), before scoring the likes of *Blade Runner* and *Chariots of Fire*.

The old Greek *rebetiko* song *Misirlou (Egyptian Woman)* has been much covered over the years. Dick Dale's rattling surf rock version was one of the standouts of Quentin Tarantino's *Pulp Fiction* soundtrack.

Playlist

Zorba's Dance
Mikis Theodorakis
Genre: Folk dance

Love Theme
Vangelis
Genre: Electronica

Secret in the Dark
Monika
Genre: Pop

The Great Dandolos
Planet of Zeus
Genre: Rock

Bloody Shadows From Afar
Lena Platonos
Genre: Electronica

Ta Pedia Tou Pirea
Melina Mercouri
Genre: Folk

Misirlou
Dick Dale
Genre: Rock

Synnefiasméni Kyriakí
Vassilis Tsitsanis
Genre: Folk

One Man's Dream
Yanni
Genre: New Age

Londra, Parisi, Athina
Sofia Vembo
Genre: Easy listening

Hungary

Read List

Eclipse of the Crescent Moon
by Géza Gárdonyi (1899, trans 1991)
The relationship of Geregely and Éva, childhood friends, later husband and wife, is charted across a backdrop of 16th-century Hungary in this novel.

The Paul Street Boys
by Ferenc Molnár (1906, trans 1927)
A group of Budapest schoolboys defend their beloved derelict corner of the city against a rival gang. This is perhaps Hungary's most internationally recognised novel.

Embers
by Sándor Márai (1942, trans 2001)
The moody parable for a Hungarian Empire on the wane, Márai's best-known work takes place in a castle hidden in the woods where an old general readies to receive a visitor he has not seen in 41 years.

Fatelessness
by Imre Kertész (1975, trans 1992)
Kertész penned this book about a Jewish Hungarian teenager's time in Auschwitz and Buchenwald concentration camps in the 1960s and 1970s. He was given Hungary's only Nobel Prize for Literature in 2002.

The Book of Fathers
by Miklós Vámos (2000, trans 2006)
Straddling 12 generations and 300 years of the trials and tribulations of the Csillag family, focusing on the male line, Vámos' magnus opus sheds rare light on Hungary's colourful but turbulent history.

Watch List

The Round-up
(1966; historic drama; directed by Miklós Jancsó)
This classic piece of World Cinema is centred around Lajos Kossuth's failed rebellion against the Habsburg Empire and the subsequent imprisonment of supporters.

Sátántangó
(1994; epic; directed by Béla Tarr)
A masterful study of the residents of a bleak village after the fall of Communism, *Sátántangó* is tragically sad, blackly humorous and cinema's most intense examination of Hungary's national character.

Sunshine
(1999; historic drama; directed by István Szabó)
Five generations of a Jewish family originally named Sonnenschein (German for 'sunshine') are followed from the Austro-Hungarian Empire to the 1956 revolution.

Control
(2003; comedy thriller; directed by Nimród Antal)
Set on a brilliantly fictionalised version of Budapest's metro, *Kontroll* is an original comedy with a dark touch in the form of a subway murderer.

Son of Saul
(2015; historic drama; directed by László Nemes)
Probably Hungary's best-known movie outside of Hungarian borders, this one depicts a day and a half in the life of Saul Ausländer, a Hungarian in the Sonderkommando (a work unit made up of prisoners in the Nazi death camps) in Auschwitz.

Liberty Bridge in Budapest; the city's transport network is the stage for the movie *Control*.

Of Liszt's 19 Hungarian rhapsodies drawing on the folk and gypsy rhythms that influenced the composer, this is the best known – a moving rendition in C-sharp minor.

The Moog became the first-ever Hungarian band to be signed by an American record label, opening doors for international fame that eluded most acts from the country. *You Raised A Vampire* is a flurry of Goth dance-rock from their second album *Razzmatazz Orfeum*. The video for it was shot cinematically in an opulently Gothic building in Budapest.

Playlist

Hungarian Rhapsody No 2
Franz Liszt
Genre: Classical

Marta's Song
Deep Forest ft Márta Sebestyén
Genre: Electro-folk

Szomorú Vasárnap
Rezső Seress
Genre: Classical

Five Hungarian Folk Songs
Béla Bartók
Genre: Classical

Add már, uram az esőt
Kati Kovács
Genre: Prog rock

You Raised A Vampire
The Moog
Genre: Indie-rock

Mielőtt elmegyek
Bikini
Genre: Rock

Maradjatok Gyerekek
Kelemen Kabátban ft Eckü
Genre: Electro-pop

Elevator
Platon Karataev
Genre: Indie

Gyöngyhajú lány
Omega
Genre: Rock

Iceland

Read List

Egil's Saga
author unknown (around 1240, trans 1893)
This classic Icelandic saga is a sprawling epic about Egil, a larger-than-life hero who spits poetry and splits skulls. Icelandic speakers can read the saga in the original Old Norse; for the rest of us, there are translation editions.

Independent People
by Halldór Laxness (1935, trans 1945)
Nobel-prize-winner Laxness's finest novel follows the life of a stubborn sheep farmer in the early 20th century, but is full of echoes of distant ages.

Jar City
by Arnaldur Indriðason (2000, trans 2004)
A thoughtful crime thriller (since turned into a movie), which begins with an old man who's been bludgeoned to death with an ashtray and probes family, genetics and organ preservation.

The Creator
by Guðrún Eva Mínervudóttir (2008, trans 2013)
This dark and intriguing novel is about a man who makes sex dolls and a mother struggling to relate to her daughter.

The Woman at 1,000 Degrees
by Hallgrímur Helgason (2011, trans 2018)
An old woman lies in bed with a hand grenade, and recalls a life that takes in South America and WWII. A bawdy, bleak and funny tour of the 20th century.

Watch List

Children of Nature
(1991; drama; directed by Friðrik Þór Friðriksson)
A couple flee their old people's home for the wilderness in a beautiful and meditative film about the countryside, and about what we do with the time we have left.

101 Reykjavík
(2000; comedy; directed by Baltasar Kormákur)
A young slacker's aimless, hedonistic world is shaken up when his mother's girlfriend comes to stay in this likeable comedy based on the book by Hallgrímur Helgason.

Of Horses and Men
(2013; drama; directed by Benedikt Erlingsson)
This award-winning account of horse breeders' relationships with each other and their steeds is a strange, watchful and beautifully shot film.

Noi the Albino
(2003; drama; directed by Dagur Kári)
An apathetic boy in the Westfjords mooches through his life until he meets a beautiful girl and starts to dream of escape in a curious, often comical drama.

Rams
(2015; drama; directed by Grímur Hákonarson)
The long-simmering conflict between two sheep farming brothers comes to a head when a cull is ordered of the diseased local flocks in an affecting, award-winning character study.

Europe

Iceland's pony breeders, and its wild landscapes, are the subject of the film *Of Horses and Men*.

Playlist

Untitled 3
Sigur Rós
Genre: Post-rock

Big Time Sensuality
Björk
Genre: Pop

Iceland's biggest musical export, Björk has trodden her own path through chart success, art pop, club culture and politics – she's also made one album (*Medúlla*) that relied almost entirely on the human voice and another (*Biophilia*) dominated by custom-made instruments.

París Norðursins
Prins Póló
Genre: Electronica

Jungle Drum
Emilíana Torrini
Genre: Pop

We Have a Map of the Piano
Múm
Genre: Electronica

Dirty Paws
Of Monsters and Men
Genre: Rock

On That Day
Ásgeir
Genre: Folk-pop

Ég er Kominn Heim
Óðinn Valdimarsson
Genre: Easy listening

Blurred
Kiasmos
Genre: Techno

As well as being one half of Kiasmos (along with Janus Rasmussen, who's from the Faroe Islands), Ólafur Arnalds is known for his scores. He has soundtracked several TV series, including *Broadchurch* and *Defending Jacob*.

Hit
The Sugarcubes
Genre: Indie

Italy

Read List

The Aeneid
by Virgil (around 19 BCE)
What the Iliad was to Ancient Greece, the Aeneid was to the Roman Republic – a national foundation myth, centred on Aeneas. The epic recounts the hero's wanderings across the Mediterranean after the fall of Troy.

The Divine Comedy
by Dante Alighieri (1320)
In this masterpiece of Renaissance literature the Florentine poet descends into the underworld with Virgil as his guide. It's both an intense religious allegory and scathing commentary on politics in medieval Italy.

The Leopard
by Giuseppe Tomasi di Lampedusa (1958, trans 2007)
A window into the Sicilian soul, The Leopard explores the life of the Prince of Salina, who struggles to maintain his influence amid the upheaval of Italian unification.

The Name of the Rose
by Umberto Eco (1980, trans 1983)
Umberto Eco's debut sees Brother William of Baskerville investigate claims of heresy in a 14th-century monastery in Northern Italy, in a novel that deftly blends philosophy with murder mystery.

My Brilliant Friend
By Elena Ferrante (2012)
Chronicling the six decade-long friendship between two girls growing up in post-war Naples, this is the first in Ferrante's celebrated quartet of Neapolitan novels.

Watch List

Bicycle Thieves
(1948; drama; directed by Vittorio De Sica)
De Sica's film follows the aftermath of a bike theft in post-war Rome – a small act with profound repercussions. Cited as one of the greatest films ever made, it is often interpreted as a parable about corruption in Italy.

La Dolce Vita
(1960; comedy; directed by Federico Fellini)
Starring Marcello Mastroianni as a young gossip journalist in Rome, *La Dolce Vita*'s depictions of hedonism, casual sex and aristocratic parties come with a dark undercurrent of existential angst.

The Battle of Algiers
(1966; drama; directed by Gillo Pontecorvo)
Star of the Italian neorealism movement, this film is a striking portrait of guerilla fighters during the Algerian War: shot newsreel style, with non-professional actors.

Cinema Paradiso
(1988; drama; directed by Giuseppe Tornatore)
Inspired by the director's own childhood, this love story follows a boy growing up in postwar Sicily and examines how his passion for his village cinema guides his life.

Gomorrah
(2008; crime drama; directed by Matteo Garrone)
Based on Roberto Saviano's non-fiction book of the same name, *Gomorrah* offers a gritty, gruesome insight into the Neapolitan mafia, the Camorra, exploring the fallout of a turf war in Naples' Scampia neighbourhood.

Mafia factions fight it out in Naples' neighbourhoods in the film *Gomorrah*.

Born in Milan to Tunisian parents, rapper Ghali switches between Italian and Arabic in politically conscious lyrics. *Cara Italia* deals with racism, his love of his homeland and – like all good Italian boys – his respect for his mother.

Originally a protest song sung by 19th-century farm workers, *Bella Ciao* went on to be the anthem of the Italian antifacist movement in WWII. In 2020, it was sung from the balconies during the first Covid-19 lockdowns.

Playlist

Europe

Azzurro
Adriano Celentano
Genre: Pop

Cara Italia
Ghali
Genre: Rap

Ma il cielo è sempre più blu
Rino Gaetano
Genre: Rock

L'Era del Cinghiale Bianco
Franco Battiato
Genre: Prog rock

Arrivederci Roma
Renato Rascel
Genre: Pop

Se Telefonando
Mina
Genre: Pop

L'Italiano
Toto Cutugno
Genre: Pop

Bella Ciao
Various
Genre: Folk

La Traviata
Giuseppe Verdi
Genre: Opera

The Ecstasy of Gold
Ennio Morricone
Genre: Soundtrack

Latvia

Read List

Among the Living and the Dead
by Inara Verzemnieks (2017)
Depicting the personal traumas and upheavals of WWII, Verzemnieks writes a poetic and deeply moving account of her grandmother's loss of home, exile and ultimate homecoming.

The Book of Riga
ed by Eva Eglaja-Kristsone & Becca Parkinson (2018)
This short story collection explores the complicated soul of Latvia's 800-year-old capital, in tales bristling with dark humor, mythic allusions and a strong sense of place.

Soviet Milk
by Nora Ikstena (2015, trans 2018)
A heartbreaking depiction of life in Soviet-occupied Latvia, in a story that explores love, motherhood, political oppression and the yearning for freedom.

Five Fingers
by Māra Zālīte (2013, trans 2017)
Latvian poet and essayist Māra Zālīte's autobiographical novel tells the 1950s story of a Latvian family leaving a Siberian prison camp and returning to their homeland after 15 years in exile.

The Cage
by Alberts Bels (1972, trans 1990)
A blend of mystery and metaphysics, Bels' page-turner is an engrossing tale about the disappearance of a well-known architect; it's also a subtle allegory of state-sanctioned repression.

Watch List

The Chronicles of Melanie
(2016; drama; directed by Viesturs Kairiss)
This biographical story about Melānija Vanaga, who was deported by the Soviets to a Siberian slave camp in 1941, is an unflinching look at Latvia's tortured past.

The Mover
(2018; drama; directed by Dāvis Sīmanis)
The suspenseful and powerful true-life story of Rīga's Zanis Lipke and his family, who risked their lives to save Jews from the Nazis.

Child of Man
(1991; drama; directed by Jānis Streičs)
Nostalgic and beautifully crafted, *Child of Man* revolves around a boy growing up in Soviet-occupied Latvia. It was the first movie ever made in Latgalian, a language spoken in eastern Latvia.

Mellow Mud
(2016; drama; directed by Renārs Vimba)
In the midst of rural poverty, 17-year-old Raya grapples with enormous obstacles to carve a new life for herself in this poignant coming-of-age story.

Bille
(2018; drama; directed by Inara Kolmane)
Based on Latvian writer Vizma Belsevica's acclaimed novel of the same name, this thought-provoking film captures Latvia's tumultuous 1930s from the perspective of a child called Bille.

In *The Mover*, the true-life tale of Latvians who saved Jews from Nazis takes place in Rīga.

One of Latvia's most beloved songs, *Saule, Pērkons, Daugava* is a stirring, triumphant work, particularly when performed by choruses that number in the thousands — as often happens during the Latvian Song Festival.

Formed in Rīga in 1979, Nāc Ārā No Ūdens (which means 'Yellow Postmen') were one of the pioneers of new wave music in the former Soviet Union, and became a cult favourite in the 1980s – ultimately influencing the entire Latvian music scene.

Playlist

Tavas Mājas Manā Azotē
Prāta Vētra
Genre: Rock

Saule, Pērkons, Daugava
Mārtiņš Brauns
Genre: Choral

Pirmais
Satellites LV
Genre: Prog rock

Ziemeļmeita
Jumprava
Genre: Synth-pop

Viņi Dejoja Vienu Vasaru
Imants Kalniņš
Genre: '60s pop

Negribas Iet Gulēt
Elizabete Balčus
Genre: Electro-pop

Nefelibata
MNTHA
Genre: Electronic

Nāc Ārā No Ūdens
Dzeltenie Pastnieki
Genre: New wave

Uguns vārdi
Oceanpath
Genre: Folk-metal

Es Nesaprotu
Rīgas Modes
Genre: Pop

Lithuania

Read List

Pavasario Balsai (The Voices of Spring)
by Jonas Mačiulis-Maironis (1895)
Writing at the threshold of Lithuania's cultural renaissance, Maironis' poetry is filled with tenderness and hope. Lithuania's identity and beauty find expression in this collection by the country's master of verse.

Forest of the Gods
by Balys Sruoga (1957, trans 1996)
A man survives the arbitrary cruelty and violence of WWII concentration camp life by using a shield of dark humour in this unflinching account, based on the author's own experiences.

Stalemate
by Icchokas Meras (1963)
The cruelty of life in Vilnius' WWII ghetto is sharply evoked in this story of a Nazi commandant who proposes to play a child prodigy at chess – and any outcome but stalemate will result in death.

Silva Rerum
by Kristina Sabaliauskaitė (2009)
The first of a four-part saga, this novel portrays a 17th-century family of nobles during the Swedish invasions. Silva Rerum is now a cultural phenomenon in Lithuania.

Vilnius Poker
by Ričardas Gavelis (1989, trans 2009)
The ceaseless anxiety and emotional erosion of life under Soviet rule saturates this philosophical novel, wracked with the protagonist's paranoid obsessions.

Watch List

Ashes in the Snow
(2018; drama; directed by Marius A Markevicius)
Teenage Lina and her family are deported from Kaunas to a Siberian gulag during WWII – but even through imprisonment and forced labour, Lina's artistic spirit remains undaunted.

Northern Crusades
(1972; drama; directed by Marijonas Giedrys)
Medieval wars come to life in this epic retelling of the life of Baltic tribal leader Herkus Mantas. Teutonic Crusaders clash with Old Prussian warriors, and courage prevails against incredible odds.

Devil's Bride
(1974; musical; directed by Arūnas Žebriūnas)
A man bargains with a demon for success, and regrets his Faustian pact too late. The village rallies against demonic trickery, but unexpected romance blooms...

Children from the Hotel America
(1990; drama; directed by Raimundas Banionis)
In this 1970s-set film, Soviet gloom fails to daunt the hippie dreams of a group of teens, who live in a reclaimed hotel and listen to a banned radio station.

The Beauty
(1969; drama; directed by Arunas Zebriunas)
Short and shocking, this film expresses ugly subjects with beauty and lyricism: the cruelty of young children and the destructive power of a single moment.

Explore beautiful Vilnius, setting for Icchokas Meras' novel *Stalemate*.

The symphonic poem *Miške*, with its enchanting woodwind and swelling strings, beautifully evokes the mystery and drama of Lithuanian nature. It remains one of the most-performed works by celebrated composer and artist Čiurlionis.

Part of the explosion of post-punk rebellion in Lithuania in the final years under the Soviet Union, Antis (a nod to 'anti-Soviet') captures the era's mood: their song *Kažkas Atsitiko* is awash with irony and disillusionment, wrapped up in anthemic ska-rock.

Playlist

Oi Šermukšnio
Ugniavijas
Genre: Traditional folk

Miške
MK Čiurlionis
Genre: Classical

Poco 1
Ganelin Trio
Genre: Jazz

Paskutinis Traukinys
FOJE
Genre: Rock

Matrica
Saulius Mykolaitis
Genre: Soft rock

Trys milijonai
Marijonas Mikutavičius
Genre: Pop/Rock

Not Afraid
GJan
Genre: Pop

Kažkas Atsitiko
Antis
Genre: Alt rock

Smėlio pilis
Saulės laikrodis
Genre: Prog rock

You Got Style
SKAMP
Genre: Pop-funk

Montenegro

Read List

The Mountain Wreath
by Petar II Petrović-Njegoš (1847, trans 1930)
Written as an epic poem in play format by the nation's most celebrated ruler, *The Mountain Wreath* is held to be a cornerstone of Montenegrin literature – despite its overt celebration of genocide.

A Tomb for Boris Davidovich
by Danilo Kiš (1976, trans 1978)
This short-story collection deals with themes of political intrigue and betrayal. Born in Serbia to a Montenegrin mother, Kiš moved to Montenegro after his Jewish father was murdered at Auschwitz.

Hansen's Children
by Ognjen Spahić (2004, trans 2011)
Written by a winner of the European Union Prize for Literature, this engrossing and affecting novel is set in Europe's last leper colony as the Romanian Communist regime collapsed around it.

The Son
by Andrej Nikolaidis (2011, trans 2012)
Written by another EU literature prize winner, *The Son* is set in the coast town of Ulcinj and follows the path of the nameless protagonist over the course of a single night.

Minister
by Stefan Bošković (2019)
The quagmire of modern Montenegrin politics is explored in this novel, as a naïve young government minister finds himself embroiled in corruption, espionage and more.

Watch List

The Battle of Neretva
(1969; action; directed by Veljko Bulajić)
Directed and co-written by Montenegro-born Bulajić, this epic WWII movie is about a real-life battle fought by the Yugoslav Partisans.

Surogat (Ersatz)
(1961; animated short; directed by Dušan Vukotić)
Wonderfully 1960s in its aesthetics, this kooky 10-minute cartoon about a day at the beach won the Academy Award for Best Animated Short Film. Vukotić lived in Croatia, but was of Montenegrin descent.

Čudo neviđeno
(1984; comedy; directed by Živko Nikolić)
In '*Unseen Wonder*', life in a small Montenegrin lakeside village is thrown into disarray by the arrival of a beautiful woman and a hare-brained scheme to connect the lake to the sea.

Packing the Monkeys, Again!
(2004; drama; directed by Marija Perović)
Montenegro's first female TV and movie director came to prominence with this debut feature about a couple struggling with their relationship while living together in a tiny apartment.

Ace of Spades: Bad Destiny
(2012; drama; directed by Draško Đurović)
Set in the 1990s, this film focuses on a former paramilitary returning to his village after being imprisoned for atrocities during the Yugoslav wars.

Europe

Rural Montenegro, a land of lakes, forests and mountains, is the backdrop to the film *Unseen Wonder*.

Playlist

To Sam Ja
Miladin Šobić
Genre: Folk

Balkan Boj
Rambo Amadeus
Genre: Hip-hop

Since his 1988 debut album, Rambo Amadeus has trodden his own eccentric path, earning him comparisons to idiosyncratic musical kooks such as Frank Zappa and Captain Beefheart. Stylistically he's hard to define, dabbling in jazz, rock, folk and, in this catchy 1989 track, hip-hop.

Džuli
Daniel
Genre: Pop

Montenegro Jazz
Perper
Genre: Rock

Crnoj Gori
Marinko Pavićević
Genre: Easy listening

Zauvijek Moja
No Name
Genre: Pop

Moj svijet
Sergej Ćetković
Genre: Easy listening

Quintet No. 4 for Guitar and Strings in D
Miloš Karadaglić
Genre: Classical

Montenegro's most successful musical export is the handsome young guitar virtuoso known simply as Miloš. He topped the UK classical chart for 28 weeks in 2011 with his debut album *Mediterráneo*, from which this track is taken.

Adio
Knez
Genre: Pop

Sonata Fantasia by Dušan Bogdanović
Montenegrin Guitar Duo
Genre: Classical

The Netherlands

Read List

The Diary of a Young Girl
by Anne Frank (1947, trans 1952)
Frank, who died in the Belsen concentration camp, spent two years filling the diary she received on her 13th birthday with entries about friendship, love and family, as well as the day-to-day trials of a life in hiding.

The Discovery of Heaven
by Harry Mulisch (1992, trans 1996)
Mulisch is one of the Netherlands' greatest writers. This ambitious novel fuses a love triangle of three troubled modern intellectuals with the efforts of an angel to locate the Ten Commandments and free God.

Why the Dutch are Different
by Ben Coates (2015)
A long-term resident of the country, Coates explores famine, floods, beer and liberalism in this travelogue of sorts, mixing humour with in-depth analysis.

The Penguin Book of Dutch Short Stories
edited by Joost Zwagerman (2016)
In this pleasingly varied collection, stories date from 1915 to the present and cover everything from colonialism to fridges with humour, sadness, surrealism and fantasy.

The Evenings
by Gerard Reve (1947, trans 2016)
A Dutch literary classic, centred on Frits van Egters, who lives a blank post-war life of routine and boredom with his parents. *The Evenings* is a gloriously downbeat book about madness and mundanity.

Watch List

Soldier of Orange
(1977; drama; directed by Paul Verhoeven)
This hard-hitting early outing for director Verhoeven and star Rutger Hauer sees a group of students divided between collaboration and resistance when the Nazis invade the Netherlands.

The Assault
(1986; drama; directed by Fons Rademakers)
A boy's family are killed in Nazi reprisals for a collaborator's murder, and he spends the rest of his life finding out the truth – and dealing with its repercussions.

Antonia's Line
(1995; comedy drama; directed by Marleen Gorris)
A widow and her daughter return home after WWII and set up a matriarchal community in this curious, tender, Oscar-winning work with a fairy-tale feel.

Wolfsbergen
(2007; drama; directed by Nanouk Leopold)
In this thoughtful, many-layered study, an old man announces his imminent death, drawing four generations of his family to gather in his country house.

The Resistance Banker
(2018; drama; directed by Joram Lürsen)
The real-life story of financier Walraven van Hall, who created an underground bank in WWII to fund the Dutch resistance, is retold in this tense, gritty drama.

Amsterdam, with its canals, coffee shops and cycling culture, exemplifies *Why the Dutch are Different.*

The Netherlands is a dance-music superpower. Megastars Tiësto, Martin Garrix and Armin van Buuren are among the highest earners in the business, packing superclubs wherever they go. The country's dance heritage also includes the ferocious early '90s genre of gabber, and the bouncy Euro dance of the Vengaboys and 2 Unlimited.

Alberts was one of the leading lights of *levenslied* ('song of life'), a popular genre that began in the 1900s. The traditional backing of accordions and barrel organs is often replaced by synths today, but the subject matter – nostalgia, love and holidays – still reflects its cheerful, escapist origins.

Playlist

Blah Blah Blah
Armin van Buuren
Genre: Trance

Radar Love
Golden Earring
Genre: Rock

1 Affoe X 2
Yung Internet
Genre: Hip-hop

Oh My God
Sevdaliza
Genre: R&B

Avond
Boudewijn de Groot
Genre: Pop

Three Dutch Folk Dances
Willem Pijper
Genre: Classical

Boom, Boom, Boom!!
Vengaboys
Genre: Euro dance

Zijn Het Je Ogen
Koos Alberts
Genre: Levenslied

The Business
Tiësto
Genre: House

Glass
Canshaker Pi
Genre: Indie

Northern Ireland

Read List

The Chronicles of Narnia
by C S Lewis (1950–56)
Lewis left Belfast for Oxford, but returned regularly. His richly readable fantasy series packs in witches, giants, talking animals and epic voyages – plus a heavy dose of Christian allegory.

North
by Seamus Heaney (1975)
Vikings, bog bodies, nature and the Troubles are among the subjects of this collection from one of the 20th century's finest and most lyrical poets.

Cal
by Bernard MacLaverty (1983)
In this sensitive account of life in the Troubles, Cal unwittingly drives an old friend to a hit on a Loyalist policeman and then falls in with the man's widow.

Dancing at Lughnasa
by Brian Friel (1990)
A man looks back on a summer in the 1930s, when his aunts hoped for love and his uncle returned from missionary work in Africa, in Friel's play about tradition, memory and change.

Milkman
by Anna Burns (2018)
A bookish 18-year-old must endure the attentions of a mysterious paramilitary leader known as 'the milkman' in a witty but creepy Booker-winning novel about power, tribalism and youth.

Watch List

Odd Man Out
(1947; thriller; directed by Carol Reed)
James Mason stars in this tense film noir about a Republican terrorist on the run from the police, who falls in love as the net tightens around him.

The Crying Game
(1992; thriller; directed by Neil Jordan)
An IRA gunman guards a kidnapped British soldier in South Armagh, then flees to London where he meets the soldier's girlfriend – and finds the IRA have not finished with him – in this haunting movie with twists aplenty.

The Boxer
(1997; drama; directed by Jim Sheridan)
Daniel Day-Lewis plays a boxer and former IRA member who returns from prison, falls for an ex and sets up a non-sectarian boxing club in Belfast, as tensions rise among fractured paramilitary groups.

Hunger
(2008; drama; directed by Steve McQueen)
Unflinching account of the 1981 hunger strike in the Maze prison, near Belfast. It's a provocative but thoughtful film about brutality and resistance.

Good Vibrations
(2013; comedy drama; directed by Lisa Barros D'Sa & Glenn Leyburn)
Joyful biopic of Terri Hooley, who fell in love with punk, opened up a Belfast record store at the height of the Troubles, and gave the Undertones their break.

Back-street Belfast is a source of limitless stories, including punk biopic *Good Vibrations*.

Belfast-born Murray was a huge star in the 1950s, and at one point in 1955 her yearning ballads took up an astonishing five spots in the UK's Top 20. She was also immortalised in Cockney rhyming slang: 'going for a Ruby', short for a Ruby Murray, meaning going to get a curry.

David Holmes ran some of Belfast's first acid house nights in the 1990s and has since forged a hugely varied career that's taken in techno, northern soul and rock. He's scored multiple Steven Soderberg movies and TV series, including *Killing Eve*, produced solo records and remixed Primal Scream and U2.

Playlist

Teenage Kicks
The Undertones
Genre: Punk

Softly, Softly
Ruby Murray
Genre: Pop

The Town I Loved So Well
Phil Coulter
Genre: Folk

Alternative Ulster
Stiff Little Fingers
Genre: Punk

Gin Soaked Boy
The Divine Comedy
Genre: Indie

Chasing Cars
Snow Patrol
Genre: Pop

Just
Bicep
Genre: Electronica

I Heard Wonders
David Holmes
Genre: Indie

Brown Eyed Girl
Van Morrison
Genre: Rock

Dance of the Blessed Spirits
James Galway
Genre: Classical

Norway

Read List

Peer Gynt
by Henrik Ibsen (1867, trans 1892)
Ibsen's epic play draws on Norway's rich folklore to tell the story of a wild young man who meets trolls in the mountains and madmen in Morocco, before returning home to face his fate.

Hunger
by Knut Hamsun (1890, trans 1899)
Future Nobel Prize winner Hamsun's angry, intense and highly influential novel follows a starving writer struggling to make his mark – and come to terms with life – in late-19th century Oslo.

The Wreath
by Sigrid Undset (1920, trans 1923)
The first of a trilogy set in rural Norwegian in the Middle Ages, this is an evocative tale of love and secrets, as a young woman fights to make her own way in the world.

The Snowman
by Jo Nesbø (2007, trans 2010)
In the seventh of Nesbø's hit Harry Hole crime novels, the troubled detective investigates a series of murders in Oslo – all linked by the snowmen found near their bodies.

A Death in the Family
by Karl Ove Knausgård (2009, trans 2012)
Knausgård's autobiographical series *My Struggle* is a literary phenomenon that's sold half-a-million copies in Norway alone. This first volume deals with his adolescence and his father's death.

Watch List

Fools in the Mountains
(1957; comedy; directed by Edith Carlmar)
Iconic Norwegian comedian Leif Juster plays a hotel manager struggling to cope with his guests, who include a film star and his ornithologist doppelgänger.

Pathfinder
(1987; drama; directed by Nils Gaup)
A boy must use his wits to survive when a murderous warband arrives in his remote village. This stunningly shot drama was the first feature-length movie to be shot in the Sami language.

Trollhunter
(2010; fantasy; directed by André Øvredal)
A group of students investigating a bear come face to face with legendary monsters, in a film of camera-jerking drama, dreamy mountainscapes and dark humour.

Kon-Tiki
(2012; drama; directed by Joachim Rønning & Espen Sandberg)
Handsome dramatisation of the Kon-Tiki expedition, in which Norwegian ethnographer Thor Heyerdahl set out to travel between South America and Polynesia on a hand-built raft.

The King's Choice
(2016; drama; directed by Erik Poppe)
In 1940, Nazi envoys pressed King Haakon VII to accept German occupation. This tense biography mixes backroom drama with furious conflict.

Both *The Wreath* novel and the film *Pathfinder* are set in rural Norway, of which there is a lot.

Traditional Norwegian music has a proud history centred around the *hardanger* fiddle (which has eight or nine strings, rather than the traditional four). In the north, traditional Sami chants, including spiritual *joik* songs, are often sung acapella or backed with drumming.

Darkthrone are one of the world's leading black metal bands. The genre was born in Norway around 1990, and is characterised by shrieked vocals, rapid drumming and raw, often high-pitched guitars, with pagan and fantasy imagery. In the 1990s the extreme end of the scene was scarred by murders and church burnings.

Playlist

In Praise of Dreams
Jan Garbarek
Genre: Jazz

Hjertebank
Ragnhild Furebotten & Tore Bruvoll
Genre: Folk

One-Armed Bandit
Jaga Jazzist
Genre: Jazz fusion

Eple
Röyksopp
Genre: Electronica

Take On Me
A-Ha
Genre: Pop

I Don't Know What I Can Save You From
Kings of Convenience
Genre: Indie

Transilvanian Hunger
Darkthrone
Genre: Metal

I Feel Space
Lindström
Genre: House

In the Hall of the Mountain King
Edvard Grieg
Genre: Classical

Don't Kill My Vibe
Sigrid
Genre: Pop

Poland

Read List

The Doll
by Bolesław Pru (1890, trans 1972)
Lauded for its realistic portrayal of all levels of 19th-century Polish society, Pru's narrative centres on his social climbing protagonist's obsessive love affair with an upper-class woman.

The Street of Crocodiles
by Bruno Schulz (1934, trans 1963)
This short-story collection showcases Schulz's magical-realist writing. Regarded as a leading literary talent of the interwar years, the Polish Jewish writer was killed during WWII.

View With a Grain of Sand
by Wislawa Szymborska (1995)
Nobel Prize laureate. Szymborska fearlessly worked through the bleak years of communism and her poetry in this volume focuses on the many wonders of daily life.

Drive Your Plow over the Bones of the Dead
by Olga Tokarczuk (2009, trans 2018)
No ordinary rural murder mystery, this novel by Booker Prize–winning Tokarczuk crams in feminism, vegetarianism and a William Blake-loving protagonist.

A Grain of Truth
by Zygmunt Miłoszewski (2011, trans 2012)
Miłoszewski's noir crime series features the world-weary prosecutor Teodor Szacki. In the picturesque town of Sandomierz, Szacki investigates ritualistic murders possibly linked to antisemitism.

Watch List

Man of Iron
(1981; drama; directed by Andrzej Wajda)
The initial rise of the Solidarity labour union movement in Gdańsk in the early 1980s is chronicled in this follow up to Wajda's 1977 feature *Man of Marble*.

Three Colours: White
(1994; drama; directed by Krzysztof Kieślowski)
The middle part of Kieślowski's triology of colour-themed movies, *White* is a dark comedy and revenge drama that skewers the freewheeling capitalism of post-communist Warsaw.

The Pianist
(2002; drama; directed by Roman Polański)
Polański returned to his native Poland to film parts of this biopic based on the memoir of pianist and composer Władysław Szpilman, one of the few survivors of the Warsaw Ghetto.

Katyń
(2007; drama; directed by Andrzej Wajda)
The massacre of some 22,000 Polish POW officers and civilians by Soviet forces in the Katyń forest during WWII is told through the eyes of the women they left behind.

Ida
(2013; drama; directed by Paweł Pawlikowski)
Winner of the Oscar for Best Foreign Language Film, this moody drama is about a novice nun in the 1960s who discovers her Jewish roots and her family's wartime past.

The Palace of Culture and Science in Warsaw is a beacon of enlightenment.

Polish-born music prodigy Chopin wrote this piece at the age of 20 and first performed it in Warsaw on 17 March 1830. Much of Chopin's music took inspiration from Polish folk and court music.

Pianist Krzysztof Komeda was one of the stars of Poland's post-war jazz scene and a notable film music composer, mainly for Roman Polanski. This 1965 work is regarded as one of the most important European jazz albums.

Playlist

Piano Concerto No 2 in F minor, Op 21
Frederic Chopin
Genre: Classical

To Ostatnia Niedziela
Mieczysław Fogg
Genre: Tango

Helokanie
Śląsk
Genre: Folk

Kalinowym Mostem Chodziłam
Aga Zaryan
Genre; Jazz pop

Stacja Warszawa
Lady Pank
Genre: Rock

Czerwone Korale
Brathanki
Genre: Modern folk

Przez Twe Oczy Zielone
Akcent
Genre: Disco polo

Astigmatic
Komeda Quintet
Genre: Jazz

Ściernisko
Golec uOrkiestra
Genre: Folk rock

Sen O Warszawie
Czesław Niemen
Genre: Rock

Portugal

Read List

The Lusiads
by Luís Vaz de Camões (1572)
Widely considered the colossus of Portuguese literature, this epic poem interweaves history and mythology in its account of the Age of Discoveries.

The Book of Disquiet
by Fernando Pessoa (1982, trans 1991)
Published posthumously, this fragmentary masterpiece by Portugal's greatest poet brings together his unpublished and unfinished works, and shines a vivid light on Lisbon.

Blindness
by José Saramago (1995, trans 1997)
This compelling, disturbing and darkly intense novel – written in a largely unpunctuated style –by Nobel Prize-winning author Saramago shows a society's descent into chaos during an epidemic of blindness.

The Maias
by Eça de Queirós *(1888, trans 1965)*
Hailed one of the greatest Portuguese works of the 19th century, this witty realist novel depicts the lives of young socialites and their wealthy families during the social and political decay of the late 1800s.

Baltasar and Blimunda
by José Saramago (1982, trans 1987)
Centred on a soldier and a girl with visionary powers, this fantastical and darkly comedic love story is set in 18th-century Portugal at the height of the Inquisition.

Watch List

Lisbon Story
(1994; drama; directed by Wim Wenders)
A love letter to Lisbon and the power of film, this beautifully shot movie tells the story of a sound engineer who journeys to Lisbon in search of a missing director.

Letters from Fontainhas
(2006; drama; directed by Pedro Costa)
A haunting, extraordinarily shot trilogy depicting immigrant lives in the impoverished slums of Fontainhas, a neighbourhood on the outskirts of Lisbon.

Night Train to Lisbon
(2013; thriller; directed by Bille August)
This Jeremy Irons movie recounts the tale of a reserved Swiss teacher who saves a woman's life and then abandons his career to embark on an intellectual adventure in Lisbon. The backdrop is Salazar's right-wing dictatorship.

The Green Years
(1963; drama; directed by Paulo Rocho)
Rocho, the master of modern Portuguese cinema, made his debut with this captivating, coming-of-age movie that deals with troubled love and shows how working-class values conflict with a bourgeois society.

Amor de Perdição
(1978; romance; directed by Manoel de Oliveira)
Tapping into the distinctly Portuguese sentiment of *saudade* (profound longing), this cinematically elegant, richly meditative four-hour movie chronicles the life of star-crossed lovers and their feuding aristocratic families.

Discover intoxicating Lisbon in the company of *The Book of Disquiet* by poet Fernando Pessoa.

Born to a poor family, the *fadista* Amália Rodrigues (1920-99) took Portuguese *fado* (folk song) from the tavern to the concert hall. The 'queen of fado', a household name in the 1940s, once famously said: 'I don't sing fado, it sings me.'

Go to a rural Portuguese festival and you'll no doubt hear the jaunty, cheeky, upbeat melodies of Pimba folk-pop, popularised by the likes of home-grown stars Quim Barreiros and Emanuel.

Playlist

Lisboa Menina E Moça
Paulo de Carvalho
Genre: Fado

Uma Casa Portuguesa
Amália Rodrigues
Genre: Fado

Haja o que Houver
Madredeus
Genre: Folk

Porto Sentido
Rui Veloso
Genre: Rock

Chuva
Mariza
Genre: Fado

Borrow
Silence 4
Genre: Pop/Rock

Os Búzios
Ana Moura
Genre: Fado

Os bichos da fazenda
Quim Barreiros
Genre: Pimba folk-pop

Para ti Maria
Xutos e Pontapés
Genre: Rock

Canção Do Mar
Dulce Pontes
Genre: Folk/Pop/Classical

IN DEPTH

Fado

from PORTUGAL

Wafting through open windows on dead-end alleys, reverberating at summer festivals and filling stadiums, fado is the folk music of Portugal's soul. Plaintive, bittersweet and emotionally intense, fado has origins that can be traced to the backstreets of working-class Alfama in Lisbon. The ditties of homesick sailors, the poetic ballads of the Moors and the bluesy, melancholic songs of Brazilian slaves have all been cited as possible influences, and no doubt fado is a blend of these and more. Fado was originally the soundtrack of the poor, the marginalized and the destitute.

Central to all forms of fado is the hard-to-translate, distinctly Portuguese concept of *saudade*, a

© JONATHAN STOKES / LONELY PLANET

© MARIO GEO / GETTY IMAGES

nostalgic longing - a relentless yearning for something absent, such as one's homeland or a distant lover. This underpins the genre's recurring themes, such as destiny (fado means 'fate'), remorse, heartbreak, unrequited love and loneliness.

Though fado is considered Portugal's national treasure – in 2011 it was added to the Unesco World Heritage list of Intangible Cultural Heritage as proof – really it is the sound of Lisbon, where it typically consists of a solo vocalist singing to the accompaniment of a 12-stringed Portuguese guitar and viola. Amália Rodrigues (1920-99) was the first to take fado to the world with her heartbreaking trills and poetic soul. More recently artists like Ana Moura and Mozambique-born Mariza have continued to broaden the genre's scope and appeal, adding a poppy beat, a pinch of blues, or a dash of tango or flamenco.

Left to right: playing fado songs in Mesa de Frades venue in Lisbon; revered singer Amália Rodrigues; modern fado artist Mariza.

Republic of Ireland

Read List

Gulliver's Travels
by Jonathan Swift (1726)
Swift's towering satire follows a surgeon who encounters flying islands, talking horses, giants and tiny people. It's surreal and disorientating, as well as angry about Ireland's condition – colonialism gets a serious kicking.

Angela's Ashes
by Frank McCourt (1996)
McCourt won the Pulitzer Prize for this often bleak, occasionally funny autobiographical tale of a 1930s Limerick family, their lives dominated by poverty and alcoholism.

Selected Poems
by W B Yeats (2000)
Yeats towers over Irish poetry and this collection gives a fine picture of the Nobel Prize winner's evolution from more traditional work to his mythic later poems.

The Sea
by John Banville (2005)
A retired art historian in southeast Ireland looks back on his life, pondering secrets, death and childhood in a wonderful, dreamlike book that won the Booker Prize.

Normal People
by Sally Rooney (2018)
Rooney's second novel follows two school friends who fall in and out of love in Sligo and Dublin. The TV adaptation was a big hit; the book is fresh, observant and compulsively readable.

Watch List

Barry Lyndon
(1975; drama; directed by Stanley Kubrick)
An 18th-century Irish rogue joins two armies and marries a rich aristocrat in Kubrick's stately, strangely detached, epic. It met with a mixed reception, but is now regarded as one of his finest.

My Left Foot
(1989; drama; directed by Jim Sheridan)
Daniel Day-Lewis won his first Oscar for his powerful portrayal of Christy Brown, a writer and painter with cerebral palsy.

The Commitments
(1991; comedy; directed by Alan Parker)
Roddy Doyle's novel about a struggling Dublin soul band is brought to joyous, bickering life in this cult 1990s crowd-pleaser.

The Wind that Shakes the Barley
(2006; drama; directed by Ken Loach)
Two brothers become comrades, then enemies, as Ireland moves towards independence in the 1920s, in an unsentimental, beautifully shot tale of revolution.

The Guard
(2011; comedy; directed by John Michael McDonagh)
Brendan Gleeson is endlessly watchable in this sly but thoughtful black comedy about an eccentric Connemara garda's bid to foil a cocaine drop.

Dublin's Trinity College Library is one of the world's most beautiful repositories of books.

Rock four-piece U2 formed in north Dublin in 1976 and have since sold around 170 million albums. They're not universally popular in their homeland – their tax affairs have met with criticism – but their tours are huge events. In the 2010s they were the only live act to gross over $1 billion.

Traditional Irish music, whether played with instruments such as fiddles, flutes, bodhrán drums and harps, or sung unaccompanied, remains hugely popular. Irish folk songs have influenced everything from US country to homegrown Celtic rock, and stars such as the Chieftains have a global profile.

Playlist

Bad
U2
Genre: Rock

Whiskey in the Jar
Thin Lizzy
Genre: Rock

A Pair of Brown Eyes
The Pogues
Genre: Folk rock

Soon
My Bloody Valentine
Genre: Shoegaze

Only Time
Enya
Genre: New Age

Something More
Róisín Murphy
Genre: Electronica

Superheroes
The Script
Genre: Pop

The Rocky Road to Dublin
The Dubliners
Genre: Folk

The Foggy Dew
The Chieftains with Sinéad O'Connor
Genre: Traditional

Boys in the Better Land
Fontaines DC
Genre: Indie

IN DEPTH

James Joyce

from REPUBLIC OF IRELAND

James Joyce is an unlikely national hero. He spent much of his life in continental Europe and wrote only three novels (two of which are very difficult to understand) and one short story collection. His poetry is not particularly popular. Yet he's a symbol of Ireland, his work celebrated in the 'Bloomsday' festival across Dublin every year.

Joyce was born in the city in 1882, and his work combines a sharp sense of place with clever, playful language. His books grew increasingly stuffed with allusions, as well as invented words such as peloothered (drunk), smilesmirk and quark (our word for a subatomic particle was first used in his experimental final novel, *Finnegans Wake*). His masterpiece, *Ulysses*, is about a

© CULTURE CLUB / GETTY IMAGES

© ANDREW MONTGOMERY / LONELY PLANET

sexually frustrated ad man mooching about a brilliantly realised Dublin, but almost every scene echoes Homer's *Odyssey*. Its final chapter contains over 20,000 words, broken into a mere eight sentences.

As well as being fiercely radical, Joyce – who is best first approached through the autobiographical *A Portrait of the Artist as a Young Man* and the short stories of *Dubliners* – has a wonderful ear for life and eye for character. By mixing the crude and everyday with the intellectual he not only changed literature but also gave a nation in flux its own modern epics. You can analyze Joyce for a lifetime, or just let his words roll off your tongue and think of Ireland.

Left to right: James Joyce; a Guinness is good for reading *Ulysses*; 16 June, the day on which *Ulysses* is set, sees costumed fans take to the streets of Dublin and beyond for Bloomsday.

Russia

Read List

Crime and Punishment
by Fyodor Dostoevsky (1866, trans 1885)
Set one sweltering summer in St Petersburg, Dostoevsky's masterpiece centres on former law student Rodion Raskolnikov and his inner world of guilt, confusion and isolation after murdering a local pawnbroker.

Kolyma Tales
by Varlam Shalamov (1989, trans 2018)
These short stories are a harrowing chronicle of the repressive gulag system in the remote Russian Far East, based on the 17 years Shamalov spent in labour camps.

Complete Poems
by Anna Akhmatova (1990)
In a country renowned for novelists, Akhmatova stands out as one of the greatest poets in the Russian language – her verses touching subjects as diverse as unrequited love, memory and the repression during Stalin's rule.

The Winter Queen
by Boris Akunin (1998, trans 2003)
The first novel in Akunin's swashbuckling detective series casts Erast Fandorin as a young policeman investigating the apparent suicide of a rich student in 1870s Moscow.

The Big Green Tent
by Lyudmila Ulitskaya (2010, trans 2015)
A vast novel acclaimed for bringing the sweep and scope of Tolstoy into the modern age, *The Big Green Tent* follows the lives of three political dissidents and how they use art to transcend the repression of their homeland.

Watch List

Battleship Potemkin
(1925; silent; directed by Sergei Eisenstein)
Eisenstein's propaganda masterpiece follows the Black Sea battleship whose crew mutinied in 1905. The massacre of civilians on the steps of Odessa is heralded as one of the most influential scenes in cinema.

Man with a Movie Camera
(1929; documentary; directed by Dziga Vertov)
Vertov's movie has no real plot, no actors, no audio. Instead we follow a cameraman as he travels through Soviet Union cities, witnessing everyday people at work and at leisure through a prism of surreal camera angles.

Come and See
(1985; war; directed by Elem Klimov)
A final flourish for Soviet cinema, the brutal, hypnotica *Come and See* stars Aleksei Kravchenko as a 15-year-old boy who joins the Belarusian partisans during WWII.

The Return
(2003; drama; directed by Andrey Zvyagintsev)
Two boys are reunited with their father after a 12-year absence and together set out to a remote island, in this masterly critique of contemporary Russian masculinity.

Leviathan
(2014; drama; directed by Andrey Zvyagintsev)
Leviathan stars the magnetic Aleksei Serebryakov as a car mechanic fighting to save his house from expropriation. Despite being part-funded by the Ministry of Culture, it was criticised at home for its bleak portrayal of rural Russia.

St Petersburg is built on as grand a scale as Dostoevsky's *Crime and Punishment*.

Vysotsky's subtly subversive songwriting delivered in a distinctive rasping voice was quietly disapproved of by the Soviet authorities in the 1960s and '70s, though tolerated on account of his immense popularity.

To coincide with glasnost and perestroika, the 1980s saw a boom in Soviet rock bands, of which Kino stood at the vanguard. Frontman Viktor Tsoi died in a car crash in Latvia in 1990 – the outpouring of grief among young fans has been likened to that for Kurt Cobain in the USA.

Playlist

Like in War
Agatha Christie
Genre: Rock

Morning Gymnastics
Vladimir Vysotsky
Genre: Singer-songwriter

What is Autumn?
DDT
Genre: Punk

Sorcerer's Doll
Korol I Shut
Genre: Horror-punk

Urban
Basta
Genre: Rap

Moscow Loves
Scriptonite
Genre: Rap

I Was Looking
Zemfira
Genre: Pop

Blood Type
Kino
Genre: Rock

Symphony No 10
Dmitri Shostakovich
Genre: Classical

Piano Concerto No 2
Sergei Rachmaninoff
Genre: Classical

Scotland

Read List

Rob Roy
by Walter Scott (1817)
Sir Walter Scott is often regarded as father of the historical novel: set at the time of the Jacobite Rising, his classic *Rob Roy* is a tale of outlaws and derring-do, against a backdrop of stirring Highland scenery.

The Prime of Miss Jean Brodie
by Muriel Spark (1961)
Set in Edinburgh, Spark's short, sparsely written masterpiece is the story of a charismatic schoolteacher, and her downfall at the hands of her favourite students.

The Living Mountain
by Nan Shepherd (1977)
Written in the 1940s, kept in a drawer until the 1970s and only gaining acclaim in recent years, Shepherd's elemental, Zen-like account of windswept Cairngorm walks is a glorious portrait of the Highlands.

Knots & Crosses
by Ian Rankin (1987)
The Edinburgh cityscape plays a starring role in Rankin's long-running Rebus series: in this first installment the inspector investigates the disappearance of young girls.

Shuggie Bain
by Douglas Stuart (2020)
The 2020 Booker Prize Winner *Shuggie Bain* is a lyrical description of growing up among the coal mining communities on the fringes of Glasgow, laced with dialogue rich in Glaswegian 'patter' (dialect).

Watch List

Culloden
(1964; docudrama; directed by Peter Watkins)
A film that applies the style of contemporary war reporting to historical events, *Culloden* recounts the defeat of the Jacobite Rising at the hands of a British army in 1746 – an event that still resonates in Scotland

The Wicker Man
(1973; horror; directed by Robin Hardy)
A police investigation in a remote Scottish island ends in pagan sacrifice in this classic horror movie.

Trainspotting
(1996; drama; directed by Danny Boyle)
Perfectly capturing the zeitgeist of the 1990s, *Trainspotting* is often considered Scotland's greatest film. The scene at the UK's highest station – Corrour on the West Highland Line – is a legendary dissection of Anglo-Scottish relations.

Ratcatcher
(1999; drama; directed by Lynne Ramsay)
The debut feature by Glaswegian director Ramsay, this unsung movie tells the story of a Glasgow childhood against the backdrop of the 1970s bin-men strike.

Sweet Sixteen
(2002; drama; directed by Ken Loach)
A teenage boy seeks a kind of redemption from a life of crime in this politically conscious drama set in Inverclyde, sometimes subtitled in English from the original Scots.

Glencoe in the western Highlands, where the Macdonald clan was massacred.

Ear-splittingly loud, Mogwai have been one of Scotland's most distinctive sounds for 25 years. Their nine studio albums go big on guitar-driven instrumentals – their debut album *Mogwai Young Team* closes with this blistering 16-minute epic.

Though not technically the Scottish national anthem, *Scotland the Brave* is musical shorthand for all things Caledonia: a 19th-century tune of obscure origins, it's played at sporting events and is blasted out by pipers in Scotland and beyond.

Playlist

Ghosts
Lau
Genre: Folk

Mogwai Fear Satan
Mogwai
Genre: Post-rock

Folk Police
Peatbog Faeries
Genre: Folk fusion

The Boy with the Arab Strap
Belle and Sebastian
Genre: Indie pop

Machines
Biffy Clyro
Genre: Rock

Bats in the Attic
King Creosote & Jon Hopkins
Genre: Folk-electronica

Idea Truck
Idlewild
Genre: Indie

Scotland the Brave
Various
Genre: Bagpipes

Nothing Ever Happens
Del Amitri
Genre: Indie

Sunshine on Leith
The Proclaimers
Genre: Folk rock

Serbia

Read List

Impure Blood
by Borisav Stanković (1910, trans 2008)
Considered Serbia's first psychological novel, *Impure Blood* tells the harrowing story of Sofka, who marries the 12-year-old son of a wealthy family to save her parents from ruin.

The Bridge on the Drina
by Ivo Andrič (1945, trans 1977)
Winner of the 1961 Nobel Prize for Literature, Serbia's most famous novel spans four centuries – from the Ottoman Era until WWI – and casts an epic eye over the complex history and relations of Serbs and Bosniaks.

Early Sorrows
by Danilo Kiš (1969, trans 1998)
This collection of 19 short stories, chronicling the end of innocence of a Jewish-Serbian village child at the dawn of WWII, is a fabulous introduction into Kiš's work.

The Houses of Belgrade
by Borislav Pekić (1970, trans 1978)
This tragicomic commentary on ethics, aesthetics and history follows Arsenie Negovan, a house-bound man who ventures out into Belgrade the day before his death.

A Guide to the Serbian Mentality
by Momo Kapor (2006)
Kapor, darling of the Belgrade literary and arts scene from the 1970s until his death in 2010, both skewers and savours the quirks of his countryfolk in this accessible peek into the psyche of Serbia and Serbians.

Watch List

Who's Singin' Over There?
(1980; black comedy; directed by Slobodan Šijan)
A filmic phenomenon following the journey of a bawdy bunch of characters on a ramshackle Belgrade-bound bus on the day before the 1941 Nazi invasion of Yugoslavia.

The Marathon Family
(1982; black comedy; directed by Slobodan Šijan)
Hectic and hilarious, this cult classic casts an absurd eye over everything from corrupt undertakers and grave-robbing gangsters to the 1934 assassination of Yugoslavia's King Alexander.

Tito and Me
(1992; comedy drama; directed by Goran Marković)
Lighthearted look at 1950s Yugoslavia through the eyes of Zoran, a chubby 10-year-old enamoured with President Tito who gets the chance to meet his hero.

Underground
(1995; black comedy; directed by Emir Kusturica)
Sprawling and surreal, arthouse auteur Kusturica's masterpiece – considered a parable of Yugoslavia's chaotic history – is fractious, brutally funny and bittersweet...and has one of the best movie soundtracks ever.

The Parade
(2011; comedy drama; directed by Srđan Dragojević)
Written following the violence of Belgrade's failed Pride parade in 2001, *The Parade* shines a spotlight on LGBT+ issues and attitudes in Serbia (spoiler alert: Serbia ain't San Fran).

Belgrade, backdrop to Borislav Pekić's tragicomic novel *The Houses of Belgrade.*

Mesečina ('*Moonlight*') – a woozy, slow-burning banger of brass and Balkan beats – is one of 11 cracking, now-classic songs off Bregović's superlatively shambolic soundtrack for iconic Serbian film *Underground.*

In a country crammed with exceptional brass musicians, the Serbian-Romani trumpeter is the best of the best. He's won so many awards – particularly at Guča, Serbia's frenetic annual trumpet festival/competition – that he now plays only as a guest of honour, to give others the chance for glory.

Playlist

Mrak
Boban Marković
Genre: Balkan brass

Mesečina
Goran Bregović
Genre: Balkan brass

Hoću Da Znam
Partibrejkers
Genre: Rock 'n' roll

Bubamara
Šaban Bajramović
Genre: Gypsy

Zlatni Papagaj
Električni Orgazam
Genre: Yugo-punk

Đurđevdan
Bijelo Dugme
Genre: Folk

Unza Unza Time
The No Smoking Orchestra
Genre: Folk/Punk

Dotako Sam Dno Života
Toma Zdravković
Genre: Folk ballad

Devojka Sa Čardaš Nogama
Đorđe Balašević
Genre: Folk rock

Rock 'n' Roll Za Kućni Savet
Riblja Čorba
Genre: Rock

Slovakia

Read List

The Year of the Frog
by Martin M Šimečka (1985, trans 1996)
At the twilight of Communism, a man grudgingly takes odd jobs after being barred from university due to his father's politics. In this semi-autobiographical novel, salvation comes through love and distance running.

The House of the Deaf Man
by Peter Krištúfek (2012, trans 2015)
Burying the past comes at a heavy cost in this piercing novel. A father-son relationship is defined by denial and repressed history – and only art can reveal the truth.

Rivers of Babylon
by Peter Pišťanek (1991, trans 2007)
At the dawn of the Velvet Revolution, the brutal gangster Rácz navigates the sleazy underworld of Bratislava, where con artists and sex workers battle for a better life.

Seeing People Off
by Jana Beňová (2008, trans 2017)
Beňová won the EU Prize for Literature with her highly observant brand of humour and tenderness displayed in this novel about a young couple living in Bratislava.

It Happened on the First of September (or Some Other Time)
by Pavol Rankov (2008, trans 2020)
In 1930s Czechoslovakia, a teenage love rivalry finds expression in a swimming contest between three friends. This award-winning novel follows them through the tumultuous decades before the Prague Spring.

Watch List

Jánošík
(1921; silent; directed by Jaroslav Jerry Siakel')
Slovakia's own Robin Hood, the folk hero Juraj Jánošík, attacks the brutal Count Šándor before joining a band of highwaymen...until the law catches up with him.

The Shop on Main Street
(1965; war drama; directed by Ján Kadár & Elmar Klos)
Tóno reluctantly accepts ownership of a Jewish shop during WWII. The real test of his courage comes when Nazis begin expelling Jews from the Slovak State.

The Copper Tower
(1970; drama; directed by Martin Hollý)
In the remote High Tatra mountains, the friendship between three men is pushed to its limits when one of their wives interrupts their retreat.

Pictures of the Old World
(1972; documentary; directed by Dusan Hanák)
Is civilisation a force for good or evil? Villagers in the Tatra Mountains are seen through a philosophical lens in this masterpiece of documentary film-making.

Candidate
(2013; thriller; directed by Jonáš Karásek)
A marketing genius bets that he can transform an unknown into a political candidate, in a thriller rife with the political scandal and unease of 2000s Slovakia.

Slovakia's Tatra mountains, the setting for the drama *The Copper Tower*.

During the Velvet Revolution, which overturned Czechoslovakia's Communist" Party, Tublatanka played this song in Bratislava's Hviezdoslavovo námestie. It became a banner song of the new era, cementing this rock anthem in Slovak history.

Marika Gombitová's soaring melodies have won dozens of awards and made her a national treasure in Slovakia. She's at her passionate best in this plaintive song, whose lyrics brim with hope and regret.

Playlist

Pravda víťazí
Tublatanka
Genre: Rock

Srdce jako kníže Rohan
Richard Müller
Genre: Pop/Rock

Architekti šťastia
Nocadeň
Genre: Alt rock

Run Run Run
Celeste Buckingham
Genre: Soul/Pop

Len bez ženy môže byť' človek blažený
František Krištof Veselý
Genre: Swing

Medzi dvoma vrchy
Zuzana Homolová
Genre: Folk

Modra
Jana Kirschner
Genre: Folk pop

Vyznanie
Marika Gombitová
Genre: Pop

Učiteľka tanca
Pavol Hammel & Prúdy
Genre: Baroque pop

Kým nás máš
Marcel Palonder
Genre: Pop ballad

Slovenia

Read List

Minuet for Guitar
by Vitomil Zupan (1975, trans 2011)
A giant of 20th-century Slovenian literature, Zupan draws on his own experiences joining the partisan fight against the Nazis in this brilliantly idiosyncratic WWII novel.

A Day in Spring
by Ciril Kosmač (1953, trans 1959)
This affecting novel recounts the narrator's return to his homeland in the aftermath of WWII following many years of absence. Kosmac's quiet story ruminates on village life, poverty, love and betrayal.

Necropolis
by Boris Pahor (1967, trans 1995)
Pahor's gripping autobiography relates his 14 months working as a medic in Nazi concentration camps. Returning decades later, he grapples with the dark memories and his overpowering feelings of survivor's guilt.

The Fig Tree
by Goran Vojnović (2016, trans 2020)
Spanning three generations, Vojnović's family saga takes readers from post-war 1950s Yugoslavia, through the Balkan wars of the 1990s and into the present.

The Tree With No Name
by Drago Jančar (2008, trans 2014)
One of Slovenia's top contemporary writers shifts between the post-communist present and the wartime occupation in this psychological novel about an archivist who uncovers a horrific past.

Watch List

Kekec
(1951; drama; directed by Jože Gale)
A much-loved icon of Slovenian cinema, this youthful adventure tale set in an idyllic mountain village revolves around a young boy who makes a daring rescue.

Outsider
(1997; drama; directed by Andrej Košak)
A Bosnian teenager moves to the Slovenian city of Ljubljana and discovers love and punk rock in this story of rebellion and political repression in 1979 Yugoslavia.

Houston, We Have a Problem
(2016; docu-fiction; directed by Žiga Virc)
Part real, part fantasy, this amusing mockumentary explores the myth of Yugoslavia's clandestine space program and touches on the cold war, the space race and philosophical questions about truth.

Cheese and Jam
(2003; comedy; directed by Branko Đurić)
This black comedy tackles prejudice and cultural stereotypes through the story of a Slovenian woman and her slacker Bosnian boyfriend, who are unwittingly drawn into a criminal underworld.

Ples v Dežju'
(1961; drama; directed by Boštjan Hladnik)
Reminiscent of French new wave cinema, critically acclaimed '*Dancing in the Rain*' mixes realism with fantastical dream sequences in a decadent tale of unrequited love.

The film *Kekec* is set in a picturesque Slovenian mountain village.

Billing themselves as the first punk rock band behind the Iron Curtain, Pankrti ('Bastards' in English) formed in 1977 and quickly gained a cult following as they ushered in a whole new era of music.

Formed in the mining town of Trbovlje in 1980, Laibach defies easy classification with its industrial rhythms, guttural vocals (often sung in German) and mocking references to totalitarianism. The group revels in controversy, even performing in North Korea in 2015.

Playlist

Ponovitev
Nesesari Kakalulu
Genre: Afrobeat

Totalna Revolucija
Pankrti
Genre: Punk

Colours
Noair
Genre: Indie pop

Zažarim
Raiven
Genre: Electropop

Zadnja Kava
Vita Mavrič
Genre: Chanson

(Rad Imel Bi) Jabuko
Katalena
Genre: Indie folk

Na Soncu
Siddharta
Genre: Alt rock

The Whistleblowers
Laibach
Genre: Avant-garde

Slovenija, Odkod Lepote Tvoje
Ansambel Bratov Avsenik
Genre: Polka

Ker Tu Je Vse Tako Lepo
Koala Voice
Genre: Indie rock

Spain

Read List

Trilogía lorquiana
by Federico García Lorca (1933–36)
This poetic and passionate trio of tragedies – *Bodas de Sangre*, *Yerma* and *La Casa de Bernarda Alba* – is a seminal work of Spain's greatest playwright, tackling love, death, revenge, yearning and social injustice in rural 1930s Spain.

Don Quixote
by Miguel de Cervantes (1605–1615)
A multilayered icon of world literature, this epic novel unveils the adventures of nobleman Alonso Quixano, who mistakenly believes himself a medieval Spanish knight.

Primera memoria
by Ana María Matute (1959, trans 2020)
A semi-autobiographical coming-of-age story of life in Mallorca in the shadow of the Spanish Civil War, by one of the most important novelists of the *posguerra* (post-war) period. It was re-translated in 2020 as *The Island*.

Patria
by Fernando Aramburu (2016, trans 2019)
Basque author Aramburu's literary sensation '*Homeland*' (turned hit TV show) dives into the devastating effects of the five-decade-long terror campaign waged in Spain by the armed Basque separatist group ETA.

Fortunata y Jacinta
by Benito Pérez Galdós (1887, trans 1986)
The great 19th-century realist's stormy four-part tale of two women, their love lives and class contrasts in 1870s Madrid is considered a Spanish masterpiece.

Watch List

Todo sobre mi madre
(1999; comedy drama; directed by Pedro Almodóvar)
A classically extraordinary Almodóvar journey of love, death, motherhood and more, paving the way for later sensations such as *Volver* with Penélope Cruz.

Spanish Affair
(2014; comedy; directed by Emilio Martínez-Lázaro)
Record-breaking comedy hit starring Dani Rovira and Clara Lago, rooted in Spanish stereotypes of the Basque Country and Andalucía.

The Sea Inside
(2004; drama; directed by Alejandro Amenábar)
Javier Bardem leads in the sensitively recreated real-life tale of Galician fisherman Ramón Sampedro, who spent three decades campaigning for the right to end his life after being left quadriplegic.

Viridiana
(1961; drama; directed by Luis Buñuel)
The great surrealist filmmaker takes on the Catholic church with this despairing, sexual, controversial saga of a pious young woman who invites a group of beggars home. It was banned upon release in Franco-era Spain.

Listening to the Judge
(2011; documentary; directed by Isabel Coixet)
From Spain's best-known female director, this Goya-award-winning documentary sees Galician writer Miguel Rivas interview former Spanish judge Baltasar Garzón about the atrocities committed by the Franco regime.

Europe

Playlist

Cositas Buenas
Paco de Lucía
Genre: Flamenco

La Leyenda del Tiempo
Camarón de la Isla
Genre: Flamenco

La Raja de tu Falda
Estopa
Genre: Rock/Flamenco/Pop

Bulería
David Bisbal
Genre: Pop

Corazón Partío
Alejandro Sanz
Genre: Pop

Lágrimas Negras
Diego el Cigala & Bebo Valdés
Genre: Flamenco/Jazz

Malamente
Rosalía
Genre: Flamenco/Pop

Amanecer
Carlos Núñez
Genre: Celtic/Folk

Qué Puedo Hacer
Los Planetas
Genre: Indie/Rock/Pop

Sobreviviré
Mónica Naranjo
Genre: Pop

The king of modern flamenco guitar, from Andalucía's Cádiz province, Paco de Lucía wows with complex bulerías, tangos, rumbas and more. He also famously collaborated with flamenco's finest Roma *cantaor* (singer), Camarón de la Isla (also from Cádiz), on 10 game-changing albums, mostly in the 1970s.

Catalonian singer-songwriter Rosalía rose to fame with her cool and contemporary, R&B-influenced flamenco creations, before hitting the international stage with collaborations alongside Pharrell Williams, J Balvin, Daddy Yankee and Maluma. *Malamente* is full of intricate handclaps, electronic beats and flamenco vocals.

Primera memoria is set in Mallorca, where Deia (above) was also the home of British poet Robert Graves.

Sweden

Read List

Pippi Longstocking
by Astrid Lindgren (1945, trans 1950)
The self-proclaimed 'strongest girl in the world' has been a rebel role model for generations of kids, with her adventures translated into more than 75 languages.

Faceless Killers
by Henning Mankell (1991, trans 1997)
Inspector Kurt Wallander's first outing illustrates Mankell's talent for combining social criticism with gripping storytelling. It doesn't shy away from uncomfortable territory including Swedish racism and national identity.

The Hundred-Year-Old Man Who Climbed out of the Window and Disappeared
by Jonas Jonasson (2009, trans 2012)
A centenarian escapes his retirement home, accidentally steals a suitcase of money belonging to a criminal gang, and is chased across Sweden in this fun, fanciful romp (later made into a film).

The Girl with the Dragon Tattoo
by Stieg Larsson (2005, trans 2008)
In the 2000s it was hard to escape this psychological thriller, the first in the mega-selling *Millennium* trilogy. It was adapted as a movie in both Sweden and Hollywood.

The Wonderful Adventures of Nils
by Selma Lagerlöf (1906-7, trans 1911)
A shrunken boy travels across Sweden on the back of a goose in this folktale-meets-geography-lesson from the first woman to win the Nobel Prize for Literature.

Watch List

Let the Right One In
(2008; horror; directed by Tomas Alfredson)
Based on a vampire novel, this is far from your typical horror thriller. It's an inventive, captivating love story about lonely teens, and it won huge acclaim upon release.

Wild Strawberries
(1957; drama; directed by Ingmar Bergman)
A road movie where the protagonist's mileage represents a journey of self-discovery. This classic offers the warmest and most accessible introduction to the canon of Bergman filmmaking.

My Life as a Dog
(1985; comedy drama; directed by Lasse Hallström)
A tender, nostalgic tale told from the perspective of 12-year-old Ingemar, searching for a place in his eccentric new home after being sent to live with relatives.

A Man Called Ove
(2015; comedy drama; directed by Hannes Holm)
It's not hard to guess the plot line when you meet Ove, an isolated, crotchety old guy looking for ways to end his life when new neighbours move in. It's a feel-good story based on a best-selling book.

Sami Blood
(2016; drama; directed by Amanda Kernell)
Escalating emotions and beautiful northern locales shine in this coming-of-age drama set in the 1930s. It serves as a stark reminder of historic attitudes to the indigenous Sami people.

In the snowy north of Sweden, the Sami people live a traditional lifestyle.

It's hard to pick a best ABBA song, but band member Frida nominates *Dancing Queen*. It's so good, she cried when she first heard it "out of pure happiness that I would get to sing that song".

Originally created to just be a concert opener, *The Final Countdown* was released in 1986 and reached number one in 25 countries. It came to be viewed as an anthem for the fall of the Berlin Wall.

Playlist

Dancing On My Own
Robyn
Genre: Dance

Dancing Queen
ABBA
Genre: Pop

The Look
Roxette
Genre: Pop

Levels
Avicii
Genre: Dance

Lovefool
The Cardigans
Genre: Pop

Hate to Say I Told You So
The Hives
Genre: Garage punk

Lush Life
Zara Larsson
Genre: Electropop

Heartbeats
The Knife
Genre: Synth-pop

The Final Countdown
Europe
Genre: '80s rock

Dance With Somebody
Mando Diao
Genre: Indie-rock

Switzerland

Read List

Heidi
by Johanna Spyri (1881, trans 1882)
This hugely popular tale of an orphan girl growing up in the mountains and moving to the city is distinguished by its cheery young heroine and lush Alps descriptions.

Steppenwolf
by Hermann Hesse (1927, trans 1929)
A middle-aged intellectual retreats into isolation and drink before meeting a beautiful woman. This dark novel about despair and healing became a countercultural classic.

The Judge and his Hangman
by Friedrich Dürrenmatt (1950, trans 1954)
A detective is found by the side of a Swiss road, a bullet in his head. Inspector Berlach, dying of cancer and nearing retirement, investigates the killing in this mystery with a philosophical bent and a sharp twist.

Man in the Holocene
by Max Frisch (1979; trans 1980)
In this much-praised meditation on life, the rains come to a Swiss-Italian village – and do not stop. An old man responds by pasting encyclopedia pages onto the walls and hiking nearby trails.

Swiss Watching: Inside Europe's Landlocked Island
by Diccon Bewes (2010)
Bewes explores the nation's quirks in a lighthearted but revealing account that takes in gun ownership, LSD and cuckoo clocks, on top of Switzerland's four languages, towering mountains and (mostly) punctual trains.

Watch List

The Swissmakers
(1978; comedy; directed by Rolf Lyssy)
A group of would-be Swiss citizens find naturalisation is not easy, in this smash-hit comedy that mixes broad humour with sly digs at Swiss quirks and bureaucracy.

Journey of Hope
(1990; drama; directed by Xavier Koller)
A rare Swiss Oscar winner, this beautifully shot film follows a Turkish family's struggle to reach Switzerland, as they deal with con men, eerie train stations and Alpine ridges on their way to a new life.

Jonah Who Will Be 25 in the Year 2000
(1976; drama; directed by Alain Tanner)
A group of acquaintances look back on the changes of the 1960s and the limitations of the '70s, in this superbly acted, Geneva-set indie classic.

My Life as a Courgette
(2016; animation; directed by Claude Barras)
This stop-motion tale of a boy who moves to an orphanage after his mother's death adroitly balances a charming account of young friendship with dark family drama.

The Divine Order
(2017; comedy-drama; directed by Petra Volpe)
In 1971, a frustrated mother in a Swiss-German farming village campaigns for women's suffrage and suggests the local women go on strike as the referendum approaches, in this appealing tale of social change.

The Swiss Alps, home to *Heidi*: harsh in winter but verdant in the summer.

The Young Gods are one of Switzerland's most influential bands. In the 1980s, they used synths, samples and pounding percussion to make urgent hard rock without a guitarist, helping shape industrial music. Trent Reznor, David Bowie and U2 have all namechecked the band.

Yodelling began in the Alps as a means of communication between peaks. Its two disciplines are *juchzin* (short yells with different meanings, such as 'it's dinner time' or 'we're coming') and *naturjodel* (where one or more voices sing a melody with no lyrics). Today, *volksmusic* (Alpine folk) acts such as Nadja Räss continue the tradition.

Playlist

Love is Not the Answer
Sophie Hunger
Genre: Indie

Skinflowers
The Young Gods
Genre: Industrial

Somebody Dance With Me
DJ BoBo
Genre: Europop

Lift U Up
Gotthard
Genre: Rock

Eisbär
Grauzone
Genre: Post-punk

Wiesel
Sonalp
Genre: Folk/World

Dr Schacher Seppli
Ruedi Rymann
Genre: Volksmusik

Nadeschka
Nadja Räss
Genre: Volksmusik

Abodah
Ernest Bloch
Genre: Classical

Oh Yeah
Yello
Genre: Electronica

Wales

Read List

The Mabinogion
authors unknown (12th century)
Warring dragons, local kings, a Roman emperor and a magical cauldron appear in this rich, influential collection of Welsh myths. It was written in medieval Welsh: one of the best translations is by Sioned Davies.

How Green Was My Valley
by Richard Llewellyn (1939)
A young man grows up in a village dominated by coal mines, the church and tradition, in a vivid, bittersweet look at late 19th-century Wales.

Collected Poems: 1934-1952
by Dylan Thomas (1952)
Thomas is as famous today for his drinking as his verse, but this collection – chosen by the poet himself – is a joy, confronting birth, death and folklore with lush language and clever rhyme.

Wales: Epic Views of a Small Country
by Jan Morris (1982)
Travel writer Morris pens a love letter to her native land, exploring its history, language, landscapes and impact on the world with lyrical enthusiasm.

Sheepshagger
by Niall Griffiths (2000)
Ianto is a buck-toothed nutcase who roams the hills while chugging drink and drugs. When his home becomes a holiday cottage, he takes revenge, in this brutal, intense and poetic novel.

Watch List

Above Us The Earth
(1977; docu-drama; directed by Karl Francis)
Francis explores the closure of a coal pit near his home in a thoughtful film that uses documentary and drama to explore the shattering effects of mining's decline in the Welsh valleys.

Human Traffic
(1999; drama; directed by Justin Kerrigan)
Five friends plan a night out, party and deal with the consequences in this exhilarating Cardiff-set cult classic of 1990s hedonism.

Submarine
(2010; comedy; directed by Richard Ayoade)
A South Wales teenager juggles young love, his troubled family, a new-age guru and school life in Ayoade's clever, touching directorial debut.

Pride
(2014; comedy; directed by Matthew Warchus)
A group of gay and lesbian activists join forces with striking miners in a funny and impassioned 1980s-set Welsh comedy drama, based on real historical events.

The Apostle
(2018; horror; directed by Gareth Evans)
This slow-building, brilliantly bizarre folk horror from the director of *The Raid* is set on a Welsh island ruled by cult leader Michael Sheen.

Travel writer Jan Morris, who died in 2020, settled in and was inspired by Wales.

Welsh alternative rock soared in the 1990s. The 'Cool Cymru' movement saw bands including the Manic Street Preachers, Stereophonics, Catatonia and Super Furry Animals scoring massive Britpop-era hits. The latter two acts sung in both English and Welsh.

Wales is *Gwlad y Gân* (the land of song). *Eisteddfods*, festivals centred on singing and poetry, have a history stretching back over a thousand years, and folk and choral music are key to national identity. Their influence runs through to modern singers such as Katherine Jenkins, Charlotte Church and Bryn Terfel.

Playlist

Delilah
Tom Jones
Genre: Pop

Motorcycle Emptiness
Manic Street Preachers
Genre: Rock

Breadfan
Budgie
Genre: Metal

Juxtaposed With U
Super Furry Animals
Genre: Indie

Diamonds are Forever
Shirley Bassey
Genre: Pop

Myfanwy
Treorchy Male Voice Choir
Genre: Folk

Home to You
Cate Le Bon
Genre: Folk rock

Hen Wlad Fy Nhadau
Bryn Terfel and the Orchestra of the Welsh National Opera
Genre: Classical

Primadonna
Marina and the Diamonds
Genre: Pop

Corner of My Sky
Kelly Lee Owens ft John Cale
Genre: Electronic

SOU AME

UTH
RICA

Argentina

Read List

Fictions
by Jorge Luis Borges (1944; trans 1962)
This anthology of fantastic tales of dreams within dreams, complex labyrinths and mysterious worlds is a good introduction to one of the most influential writers of the 20th century.

The Tango Singer
by Tomás Eloy Martínez (2004, trans 2006)
In this atmospheric Buenos Aires novel, a student becomes obsessed with decoding Borges and enthralled in the quest to find an elusive tango singer.

Kiss of the Spider Woman
by Manuel Puig (1976, trans 1979)
Later adapted for the screen and stage, this novel depicts a series of dialogues between two cellmates in an Argentine prison who develop a strong bond.

Hopscotch
by Julio Cortázar (1963; trans 1966)
Cortázar's experimental novel is prefaced by a table of instructions laying out a non-chronological sequence in which the chapters can be read, allowing the reader to hopscotch their way through.

The Tunnel
by Ernesto Sábato (1948, trans 1988)
A dark, existential novel (admired by French philosopher Albert Camus) in which a jailed artist candidly narrates the chilling tale of the events that led to his imprisonment, while remaining unrepentant.

Watch List

The Official Story
(1985; drama; directed by Luis Puenzo)
During the final year of the military dictatorship of 1976-1983, a woman begins to realise the extent of the regime's brutality and her own unknowing complicity.

Nine Queens
(2000; drama; directed by Fabián Belinsky)
Two con artists scam their way around the streets of Buenos Aires in a film brimming with humour that depicts a city full of tricksters.

The Headless Woman
(2008; drama; directed by Lucrecia Martel)
Set in Salta, this unsettling film about a woman left confused after hitting something with her car conveys themes of bourgeois repression, guilt and denial.

The Secret of their Eyes
(2009; drama; directed by Juan José Campanella)
Oscar-winning thriller in which a former prosecutor writes about a gruesome murder case he worked on 25 years earlier, forcing him to revisit the past.

The Man Next Door
(2009; comedy drama; directed by Mariano Cohn & Gastón Duprat)
The story of the increasingly fractious relationship between a seemingly reasonable man and his uncouth neighbour, set in the famous Le Corbusier-designed Casa Curutchet building in La Plata.

Buenos Aires is the best place to encounter the controlled passion portrayed in *The Tango Singer*.

Composer Astor Piazzolla (1921-1992) incorporated elements of jazz and classical music into his work, creating a new form of tango. In this recording, pianist Daniel Barenboim is accompanied by *bandoenist* Rodolfo Mederos and double bassist Hector Console.

Mercedes Sosa (1935-2009) was a folk singer from Tucumán who recorded songs written by many different South American artists. One of the most popular is *Alfonsina y el Mar*, which pays homage to Argentine poet Alfonsina Storni.

Playlist

De Música Ligera
Soda Stereo
Genre: Rock

No Me Arrepiento de Este Amor
Gilda
Genre: Cumbia

Todas las Hojas Son del Viento
Pescado Rabioso
Genre: Rock

Piazzolla: Otoño Porteño
Daniel Barenboim
Genre: Tango

Por Una Cabeza
Carlos Gardel
Genre: Tango

Alfonsina y el Mar
Mercedes Sosa
Genre: Folk

Vasos Vacíos
Los Fabulosos Cadillacs
Genre: Ska

Zamba para Olvidarte
Daniel Toro
Genre: Zamba

Tuve que Quemar
Sara Hebe
Genre: Hip-hop

Rezo por Vos
Charly García
Genre: Rock

Bolivia

Read List

American Visa
***by Juan de Recacoechea* (1994, trans 2007)**
While a teacher awaits a US visa in La Paz, he spins a vivid yarn about the Bolivian capital that takes in dive bars, corrupt politicians and what just might be love.

The Fat Man from La Paz: Contemporary Fiction from Bolivia
edited by Rosario Santos (2000)
Twenty short stories cover everything from the Chaco War of the 1930s to crime and sexuality, offering a rare dive into Bolivian literature in English.

Marching Powder
by Rusty Young (2003)
This Australian author's hit book is based on the true story of a British-Tanzanian man who tried to smuggle 5kg of cocaine out of La Paz airport, and ended up in a prison with an in-house cocaine factory and a primary school.

Norte by Edmundo Paz Soldán
(2011, trans 2016)
Paz Soldán is one of the leading lights of McOndo, a Latin American genre that gives magic realism an urban twist. Norte presents the stories of a serial killer, a student and an artist in this thrilling border tale.

Our Dead World
by Liliana Colanzi (2016, trans 2017)
This playful short-story collection from one of Bolivia's most talented contemporary writers takes in slavery, suicide, folklore and faraway planets.

Watch List

Blood of the Condor
(1969; drama; directed by Jorge Sanjinés)
An NGO that purports to help Indigenous communities is actually sterilising women, in this passionate drama based on real testimonies. It was filmed in Quechua, Spanish and English.

Chuquiago
(1977; drama; directed by Antonio Eguino)
This compelling study of La Paz life unfolds around an Indigenous boy, a teenager high school drop-out, a bureaucrat and a young woman with a marriage proposal.

The Secret Nation
(1989; drama; directed by Jorge Sanjinés)
Mixing a highland pilgrimage with flashbacks, this film centres on an Aymara coffin maker who returns to the village he was born in to dance to the gods and die.

The Devil's Miner
(2005; documentary; directed by Kief Davidson & Richard Ladkani)
Cerro Rico, near Potosí, is rich with silver and home to brutal working conditions. This thoughtful documentary follows two teenage miners into the belly of the mountain.

Southern District
(2009; drama; directed by Juan Carlos Valdivia)
Valdivia puts a wealthy white Bolivian family and their Aymara servants under the microscope in this moving domestic drama, at a time when Evo Morales' government was working to transform Indigenous rights

La Paz offers a world of stories, whether you visit this high-altitude city or experience it from afar.

Folklorica (Bolivian folk), uses instruments such as the *charango* (an Andean lute), *siku* (pan pipes), flutes, rattles and percussion. Different dances have their own backing, such as *morenada* (based on African slave stories), *tobas* (admired by the Inca) and *kullawada* (associated with La Paz).

Born in 1949, Luzmila Carpio sings mostly in Quechua, a choice that initially met some resistance from a snobbish establishment. She has become a star and an advocate for Indigenous rights. Her fluttering vocals found a new global audience after a collaboration with Argentine electronica label ZZK.

Playlist

Hoja Verde
Atajo
Genre: Ska-rock

Sonata Chiquitanas
Florilegium, Gian-Carla Tisera, Alejandra Wayar, Henry Villca, Katia Escalera
Genre: Classical

Llorando Se Fue
Los Kjarkas
Genre: Folklorica

Aires de Tomoyo
Celestino Campos
Genre: Folklorica

K'oli Pankarita
Zulma Yugar
Genre: Folklorica

Ch'uwa Yaku Kawsaypuni
Luzmila Carpio
Genre: Folklorica

Carahuata
Los Jairas
Genre: Folklorica

Suelta Todo
Sibah
Genre: Rock

Nadie Muere de Amor
Huáscar Aparicio
Genre: Folklorica

Inkas
Grupo Aymara
Genre: Folklorica

South America

Brazil

Read List

The Posthumous Memoirs of Brás Cubas
by Machado de Assis (1881, trans 1997)
In this avant-garde novel, a deceased author writes about the most important moments of his life, resulting in a unique critique of late 19th-century Brazilian society.

The Slum
by Aluísio Azevedo (1890, trans 2000)
A naturalistic novel denouncing the exploitation and terrible living conditions of the residents of Rio de Janeiro's tenements at the end of the 19th century.

Barren Lives
by Graciliano Ramos (1938, trans 1965)
One of the great classics of Brazilian literature, this novel portrays the miserable life of a family of refugees punished by drought: the Brazilian *Grapes of Wrath*.

The Hour of the Star
by Clarice Lispector (1977, trans 1992)
This novel shares the coming-of-age story of Macabéa, an impoverished 19-year-old girl living in Rio de Janeiro whose life meets a sudden and tragic end.

The Brazilian People
by Darcy Ribeiro (1995, trans 2000)
This modern classic, written by one of 20th-century Brazil's most influential anthropologists, reviews the full history of the formation of Brazilian civilisation in an eloquent and dramatic fashion.

Watch List

Bye Bye Brasil
(1980; drama; directed by Caca Diegues)
Three travelling artists drive from the Amazon to Brasilia in an RV, performing shows for the most humble sector of the Brazilian population, many of whom still do not have access to television.

Central Station
(1998; drama; directed by Walter Salles)
A bitter former teacher who makes a living writing letters for illiterate people joins a young boy on a healing trip through the Northeast of Brazil to find his father.

A Dog's Will
(2000; comedy; directed by Guel Arraes)
This film chronicles the adventures of João and Chicó, two poor men who steal from a small Brazilian village to survive but are saved by an apparition of the Virgin Mary.

City of God
(2002; crime/drama; directed by Fernando Meirelles & Kátia Lund)
In this highly acclaimed, Oscar-nominated film, a poor, young black man growing up in Rio's most violent *favela* (slum) analyses daily life through his talent for photography.

Elite Squad
(2007; action; directed by José Padilha)
A police captain in Rio de Janeiro is exhausted and about to retire, but he needs to find a successor to lead a dangerous mission.

In Rio de Janeiro, rich and poor share a stunning natural theatre.

Girl from Ipanema, as it's titled in English, is a classic Brazilian bossa nova and jazz song – bossa nova being a style of samba that came about in the 1950s and '60s in Rio de Janeiro. The track won a Grammy for Record of the Year in 1965 and was a global hit that continues to unite and inspire today.

This rousing rock hit from Legião Urbana is a Brazilian staple and one of the first songs to be labelled 'Brazilian Rock.'

Playlist

Sampa
Caetano Veloso
Genre: Pop

Águas de Março
Various artists
Genre: Bossa nova

Ó Abre Alas
Chiquinha Gonzaga
Genre: Folk

Brasil
Cazuza
Genre: Pop/Rock

Garota de Ipanema
Tom Jobim
Genre: Bossa nova

Asa Branca
Luiz Gonzaga
Genre: Choro

Isto Aqui, o Que é?
Ary Barroso
Genre: Pop

Que País é Esse?
Legião Urbana
Genre: Brazilian rock

País Tropical
Jorge Ben Jor
Genre: Samba rock

Aquarela do Brasil
Gal Costa
Genre: Samba rock

Chile

Read List

Twenty Love Poems and a Song of Despair
by Pablo Neruda (1924, trans 1969)
Neruda won the Nobel Prize in Literature in 1971; this collection of lush but measured romantic poems has sold over a million copies.

The House of the Spirits
by Isabel Allende (1982, trans 1985)
Allende's bestselling, magic-realist epic follows a family through three generations in an unnamed South American country, mixing love, hope and violence, and moving from a rural hacienda to a military coup.

Clandestine in Chile
by Gabriel García Márquez (1986, trans 1987)
The revered Mexican novelist tells the story of Miguel Littín, a real-life Chilean exile who returned to the country undercover to shoot a film exposing the truth about Augusto Pinochet's dictatorship.

By Night in Chile
by Roberto Bolaño (2000, trans 2003)
Bolaño is one of the great Latin American novelists. This, his first work to be published in English, is a priest's dying account of his life as a literary critic and instructor to Pinochet; a fierce satire on Chile and its literary scene.

My Tender Matador
by Pedro Lemebel (2001, trans 2005)
An ageing man is befriended by a handsome young revolutionary in this funny and touching novel that mixes love, dictatorship and *bolero* music.

Watch List

Three Sad Tigers
(1968; drama; directed by Raúl Ruiz)
Ruiz directed more than 100 films in his 40-year career, winning numerous awards. His black-and-white debut follows a few wild, boozy days in the life of three Santiago residents, as the threat of violence draws closer.

Machuca
(2004; drama; directed by Andrés Wood)
This film, about a boy from a wealthy but dysfunctional family in the tense final days of Salvador Allende's socialist government, broke Chilean box-office records.

Gloria
(2013; drama; directed by Sebastián Lelio)
A middle-aged divorcee (played by an excellent Paulina Garcia) dives back into Santiago's dating scene with chaotic results, in this sharp but sympathetic comedy.

No
(2012; drama; directed by Pablo Larraín)
In 1988, Chile held a referendum on whether dictator Pinochet should stay in power. This darkly funny, historical drama features Gael García Bernal as an ad man promoting the 'no' campaign in the face of censorship and intimidation.

A Fantastic Woman
(2017; drama; directed by Sebastián Lelio)
A transgender woman's life falls apart after her boyfriend dies in this brilliant, moving study of prejudice and grief. It won the Best Foreign Language Film Oscar.

Author Isabel Allende was granted the US Presidential Medal of Freedom in 2015 by President Obama.

Mon Laferte is one of Chile's biggest stars, whose songs achieve streaming figures in the hundred millions. Born in Viña del Mar, she got her break in a 2003 Chilean talent show. Since then, she's recorded everything from Latin pop to *bolero* and heavy metal.

The *cueca* is Chile's national dance. It has its roots in the *zamacueca* dance of Peru (itself a variant of the Spanish *fandango*), which was popular in taverns across 19th-century Chile. Traditionally, the male dancer's moves echo the strut of an eager cockerel and the woman's a more demure hen.

Playlist

Tu Falta de Querer
Mon Laferte
Genre: Latin pop

Mentira
La Ley
Genre: Rock

Trees
Föllakzoid
Genre: Rock

Hablar de Ti
Gepe
Genre: Latin pop

El Guatón Loyola
Los Perlas
Genre: Cueca

Qué Dirá el Santo Padre
Quilapayún
Genre: Cueca

Papi ¿Donde Está El Funk?
Los Tetas
Genre: Funk rock

Y.G.H.
Ricardo Villalobos
Genre: House

Entonada
La Marraqueta
Genre: Jazz fusion

Santiago
Newen Afrobeat
Genre: Afrobeat

South America

Colombia

Read List

María
by Jorge Isaacs (1867, trans 1890)
One of the definitive works of 19th-century Spanish Romantic literature, Isaac's only novel depicts the tragic story of lovebirds María and Efraín, set in Valle Del Cauca.

Love in the Time of Cholera
by Gabriel García Márquez (1985, trans 1988)
This famous novel by Nobel Prize winner Márquez follows the epic love story of Florentino Ariza and Fermina Daza, set in an unnamed Colombian port city suggested to be Cartagena.

Our Lady of the Assassins
by Fernando Vallejo (1994, trans 2001)
This semi-autobiographical novel garnered Vallejo international attention when it was adapted into a film, which led to the novel's translation into English. Set in Medellín, it's the story of a writer called Fernando who meets and fall in love with local teen hitman Alexis.

The Vortex
by José Eustasio Rivera (1924, trans 2003)
Set in more than three of Colombia's bioregions, including the glorious rainforest and eastern plains, Rivera's novel follows the elopement of adventurer Arturo Cova and his lover Alicia.

In the Beginning Was the Sea
by Tomás González (1983, trans 2014)
Based on a true story, this satirical novel follows a young couple from Medellín as they abandon their luxurious city life to settle down on a remote island and be closer to nature; a move that turns into a disaster.

Watch List

Addictions and Subtractions
(2004; drama; directed by Víctor Gaviria)
In this fast-paced drama, a real-estate agent looking for a get-rich-quick scheme begins trafficking cocaine and his life quickly devolves into a world of chaos and violence.

The Strategy of the Snail
(1993; comedy drama; directed by Sergio Cabrera)
A Berlin International Film festival winner, this tragicomedy tells the story of a group of low-income families squatting in a Bogota mansion who fight greedy realtors to keep their home.

The Mansion of Araucaima
(1986; horror; directed by Carlos Mayolo)
In this gothic horror-drama a young model runs away from the set of a commercial she's filming and enters the mansion of Araucaima, where its dwellers indulge in strange rites.

Impunity
(2011; documentary; directed by Juan José Lozano & Hollman Morris)
This chilling film documents the atrocities committed by Colombian paramilitary forces through a series of legal hearings witnessed by the victim's friends and family.

The Wind Journey
(2009; drama; directed by Ciro Guerra)
A vallenato musician travels thousands of miles to return his accordion to his original teacher in this moving story filmed across 80 locations in northern Colombia.

Gabriel García Márquez worked as a journalist in Cartagena (above).

Bomba Estéreo has been making waves since their funky, eclectic band was formed in Bogotá in 2005. This *Fuego* single, a dynamite crossover between psychedelic pop and hip-hop, hit the airwaves in 2011.

Hailing from Mompox in the Bolivar province, Toto's hit 1993 album *La Candela Viva* was a platinum-selling phenomenon. Her spirited Afro-Caribbean music is such a Colombian staple that during the Nobel Prize ceremony in 1982 she accompanied Gabriel García Márquez as part of the Colombia cultural delegation.

Playlist

Cumbia Sampuesana
Aniceto Molina
Genre: Folk-pop

Real
La Etnia
Genre: Hip-hop

La Papaya
Noel Petro
Genre: Folk

Fuego
Bomba Estéreo
Genre: Pop/hip-hop

Florecita Rockera
Aterciopelados
Genre: Rock

La Candela Viva Toto la Momposina
Totó la Momposina
Genre: Folk

Alicia Adorada
Alejo Duran
Genre: Folk

Mas Allá del Dolor
Masacre
Genre: Heavy metal

Oiga, Mire, Vea
Guayacán Orquesta
Genre: Pop

Busco Alguien Que Me Quiera
El Afinaito
Genre: Pop

Ecuador

Read List

Cumandá: The Novel of the Ecuadorian Jungle
by Juan León Mera (1887, trans 2007)
In this classic tale of star-crossed lovers, Carlos, the son of a rancher-turned-friar, and the stunningly beautiful Cumandá from a tribe in the Ecuadorian Amazon, suffer a tragic demise.

Huasipungo
by Jorge Icaza (1934, trans 1962)
This brutally realistic novel tells of a greedy property owner who exploits and brutalises the indigenous Ecuadorians in his community to build his wealth.

Wolves' Dream
by Abdón Ubidia (1986, trans 1997)
In this fast-paced novel by award-winning fiction writer Ubadia, five characters looking for quick cash in 1980 Quito devise a plot to rob a bank– to devastating results.

The Devil's Nose
by Luz Argentina Chiriboga (2010, trans 2015)
A prominent voice for the Afro-Ecuadorian community, Chiriboga's dramatic novel details the experience of thousands of Jamaicans who immigrated to Ecuador in the late 1800s searching for a better life.

When The Guayacanes Flourished
by Nelson Estupiñán Bass (1950, trans 1987)
This unique novel explores a famed moment in Ecuadorian history with the retelling of the assassination of Eloy Alfaro, a leader of Ecuador's liberal revolution in the early 1900s.

Watch List

Ratas, Ratones, Rateros
(1999; drama; directed by Sebastián Cordero)
A thief named Salvador has his life turned upside down when his cousin, an ex-convict searching for a place to hide and fast cash, drags him into his criminal underworld.

With My Heart in Yambo
(2011; documentary; directed by Maria Fernanda Restrepo)
Director Restrepo investigates the 1988 disappearance of her two older brothers who were tortured and murdered by police, to devastating effects on her family.

How Much Further
(2006; drama; directed by Tania Hermida)
Award winning Ecuadorian filmmaker Hermida films the story of a Spanish tourist and an Ecuadorian student who take bus rides together when roads are blocked due to indigenous strikes.

A Tus Espaldas
(2011; drama; directed by Tito Jara)
In '*Behind Your Back*', a bank teller in Quito lives out his insecurities as a mestizo son of his migrant mother by overspending to impress people on the 'right' side of town.

Alba
(2016; drama; directed by Ana Cristina Barragán)
In Barragán's award-winning first film, 11-year old Alba moves in with her father whom she hardly knows when her mother becomes ill. Though they both want love, what follows is a poignant struggle to connect.

South America

Playlist

Nuestro Juramento
Julio Jaramillo
Genre: Folk

Yo Nací Aquí
Juan Fernando Velasco
Genre: Pop

El Aguacate
Juan Fernando Velasco
Genre: Pop

Dicen
Pamela Cortes
Genre: Pop

Tanto Ganas Tanto Pierdes
Verde70
Genre: Rock

With an opening electric guitar hook that just won't quit, and just the right amount of Colombian rock influence, this jamming rock ballad from Verde70 is the perfect example of contemporary South American rock at its finest.

A Tajitos De Caña
Hernán Sotomayor
Genre: Folk

This nostalgic folk song recently celebrated its 45th anniversary as an Ecuadorian cultural classic in 2020. Written by famous folk songwriter Sotomayor when he was only 17 years old, it was inspired by his fond memories of adolescence and young love while growing up on the reeded riverbanks of Valle Hermoso, in Catamayo.

A Mi Lindo Ecuador
Fernando Pacheco ft Rubén Barba
Genre: Rock

Light It Up
Esto es Eso
Genre: Pop

Sometimes Ok
Fausto Miño
Genre: Pop

Kito con K
Sal y Mileto
Genre: Rock

Ecuador's Amazon is the setting for Juan León Mera's romantic novel *Cumandá*.

French Guiana

Read List

Papillon
by Henri Charrière (1969, trans 1973)
A smash success upon its publication in France, this loosely autobiographical work describes life in French Guiana's most notorious penal colony, culminating with French gangster Charrière's daring escape in 1944.

Bozambo's Revenge
by Bertène Juminer (1968, trans 1976)
Cayenne-born doctor and writer Juminer flips the colonial paradigm and makes black Africans the superior dominant empire, colonising France and revising its cultural landmarks in this deliciously satirical novel.

Space in the Tropics
by Peter Redfield (2000)
The surprising intersection of nature and culture is on show in this pioneering anthropological work, which encompasses tropical medicine, penal colonies, Robinson Crusoe, ecotourism and outer-space exploration.

The Governor's Daughter
by Paule Constant (1994, trans 1998)
Set in Cayenne's French penal colony after WWI, a young girl finds more humanity among the prisoners than in her austere parents in this comedic but disturbing tale.

Travels with Tooy
by Richard Price (2008)
Noted American scholar Price introduces the visionary world of Tooy – a philosopher, healer, priest, oral historian and unrivaled storyteller living on the edge of Cayenne.

Watch List

The Pure Life
(2014; drama; directed by Jeremy Banster)
Filmed in French Guiana's dense jungles, Banster's award-winning biopic of explorer Raymond Maufrais captures the excitement and danger of travelling into the unknown.

Le Bouillon d'Awara
(1996; documentary; directed by Cesar Paes)
Using the rich national stew of *awara* as a way in, this film celebrates the multicultural nature of French Guiana, including Amerindians, Brazilians, Laotians and African descendants.

Maroni, Les Fantômes du Fleuve
(2018; drama; directed by Olivier Abbou)
Capturing the beauty and mystery of the Guyanais rain forest, this imaginative series revolves around a group of detectives drawn into a string of frightening cases.

Jean Galmot, Aventurier
(1990; drama; directed by Alain Maline)
The inspiring story of French journalist Jean Galmont, who moved to French Guiana in search of gold and became a champion for the country's independence.

Papillon
(1973; drama; directed by Franklin J Schaffner)
Steve McQueen stars in this thrilling prison drama based on a true story – adapted from the book by Charrière (see left) – which shows the horrors and brutality inside France's most dreadful penal colony.

South America

Playlist

Le Bal Masqué
La Compagnie Créole
Genre: Pop

Ba Mo Mo Ano Doré
Loonah et les Frères Cippe
Genre: Traditional

Ancrée à Ton Port
Fanny J
Genre: R&B

La Revanche des Tololos
Les Mécènes
Genre: Carnavalesque

Jardin d'Hiver
Henri Salvador
Genre: Chanson

Aleké
Les Bushinengé
Genre: Traditional

A Tes Côtés
Tina Ly ft Richard Cavé
Genre: Zouk

Dweet So
Jahyanai
Genre: Dancehall

Validé
Cween
Genre: R&B

Kalkilé Mo Ka Kalkilé
Ann'Klod
Genre: Traditional

Over a 70 year artistic career, Henri Salvador (1917–2008) made a profound impact on world music. He played with Django Reinhardt, recorded some of the first French rock-and-roll songs, contributed to Brazilian bossa nova and even created some of the first music videos.

The Bushinengé are descendants of enslaved Africans who escaped Surinamese plantations and established communities in the forest. The songs feature heavily syncopated drum rhythms and pay homage to nature deities or deceased ancestors.

There were several French penal colonies in French Guiana, one of which played a part in classic flick *Papillon*.

Guyana

Read List

Palace Of The Peacock
by Wilson Harris (1960)
Set in the 16th century, a group of men from different ethnic backgrounds make their way up a treacherous river in the jungles of Guyana. Harris' narrative takes readers on a journey to a dark period in the nation's history.

Poems of Resistance from British Guiana
by Martin Carter (1954)
The Guyanese poet and political activist's collection of poems, which reflected the tragedy and hope of 1950s Guyana, was published following his (first) stint in jail for 'spreading dissension'.

Kaywana trilogy
by Edgar Mittelholzer (1952-58)
Children of Kaywana, *Kaywana Stock* and *Kaywana Blood* together form an iconic saga that, while fictional, provides a solid history lesson on Guyana through the ages.

The Ventriloquist's Tale
by Pauline Melville (1997)
Two illicit love affairs are the vertebrae of this absorbing novel set against the background of colonial life in Guyana over three generations.

Coolie Woman: The Odyssey of Indenture
by Gaiutra Bahadur (2013)
The Guyanese-American writer traces her great-grandmother Sujaria's journey from India to work as an indentured labourer on a sugarcane plantation in British Guiana, in this multi-award-winning biography.

Watch List

Antiman
(2014; drama; directed by Gavin Ramoutar)
In this award-winning short exploring LGBT issues, a young Guyanese boy must win a cricket tournament to attend the local masquerade and see the boy he pines for.

The Seawall
(2010; drama; directed by Mason Richards)
Another award-winning short, charting the emotional journey of a boy from his home in Guyana to a new life in Brooklyn – a move that the filmmaker himself made at the age of seven.

Jungle Fish
(2012; documentary; directed by Louisiana Kreutz)
Outlining the struggle for economic independence by the Indigenous people of Guyana, three fishermen voyage into the jungle on a mission to prove that the country's fledgling sport fishing industry is viable.

Thunder in Guyana
(2003; documentary; directed by Suzanne Wasserman)
This docu-film recounts the love story between Chicago-born Janet Rosenberg and Guyanese Cheddi Jagan, who set off for the British colony to start a socialist revolution that led them first to jail and later to the presidency.

Adero
(2017; fantasy, drama; directed by Kojo McPherson)
Vivid, haunting dreams push a Guyanese man of Amerindian and African heritage to search for the identity of his parents in this surreal short film.

Jungle Fish offers a portrait of the indigenous communities in Guyana's Amazon region.

A founding member of The Equals, one of the UK's first racially integrated pop groups, Guyanese-born Eddy Grant's subsequent solo hit *Electric Avenue* went platinum in the UK.

Bursting onto Guyana's soca scene in 2017, the vibrant young artist known as Blaze Anthonio (real name Marlon Ashford Simon) is known for his songs that advocate for social change in Guyana.

South America

Playlist

Guyana With Love
Keith Waithe
Genre: Jazz/Folk

Guyana Baboo
Terry Gajraj
Genre: Chutney

I Won't Stop
Timeka Marshall
Genre: Soca

Electric Avenue
Eddy Grant
Genre: Pop/Reggae

Cricket In The Jungle
Dave Martins & the Tradewinds
Genre: Calypso/Pop/Folk

So Lucky in My Life
Telstars ft Aubrey Cummings
Genre: Pop

Seh No
Blaze Anthonio
Genre: Soca

Sound Odyssey
Loyle Carner & Keith Waithe
Genre: Folk/Rap

Second Chance
Mark Batson
Genre: Reggae

It Burns Inside
Johnny Braff
Genre: Rock/Pop

Peru

Read List

Conversation in the Cathedral
by Mario Vargas Llosa (1969, trans 1975)
Two men under the shadow of Peru's 1950s dictatorship meet in a dog pound and talk in a bar, in this searing, challenging exploration of power and corruption from one of Latin America's most influential writers.

Selected Poems
by César Vallejo (2006)
Peru's greatest poet wrote often sad, sometimes surreal modernist verse about everything from his own imprisonment in Peru to the Spanish Civil War (which he witnessed first hand).

Red April
by Santiago Roncagliolo (2006, trans 2010)
A Peruvian prosecutor explores a gruesome series of killings that may be connected to the Shining Path terrorist organisation, in this thoroughly engaging thriller.

Turn Right at Machu Picchu
by Mark Adams (2011)
In this travelogue, Adams tramps to what he drolly describes as 'the lost summer home of the Incas' while telling the stories of the great Andean civilisation and of Hiram Bingham, who 'discovered' the site in 1911.

The King is Always Above the People
by Daniel Alarcón (2018)
Journalist and podcaster Alarcón's varied and endlessly curious collection of short stories explores big topics like migration and war with a light touch.

Watch List

Aguirre, the Wrath of God
(1972; drama; directed by Werner Herzog)
This hallucinatory tale of conquistadors marching down the Andes and into the Amazon in search of gold is a frenzied tour-de-force, starring Klaus Kinski as a gaunt madman driving the invaders onward to destruction.

The City and the Dogs
(1985; drama; directed by Francisco J Lombardi)
This award-winning adaptation of Mario Vargas Llosa's *The Time of the Hero* tells the story of a young cadet who witnesses a murder, and explores individual choice and institutional cruelty.

Madeinusa
(2006; drama; directed by Claudia Llosa)
It's festival-time in a remote Quechua village and a young woman dreams of leaving, in this unsettling tale about tradition and change.

Undertow
(2009; drama; directed by Javier Fuentes-León)
A fisherman is visited by the ghost of his lover, who drowned at sea and wants to know where his body is, in an appealing piece of magic realism.

When Two Worlds Collide
(2016; documentary; directed by Heidi Brandenburg Sierralta & Mathew Orzel)
In 2009, violence broke out in the Amazon over the opening up of tribal lands to mining and drilling. This striking documentary takes the view of the underdog.

In *Turn Right at Machu Picchu*, writer Mark Adams relates the exploits of American adventurer Hiram Bingham.

Singer Susana Baca is the biggest voice in Afro-Peruvian music. She's toured around the world, served as Peru's Minister of Culture and is a charismatic live performer who educates as she entertains.

Much Peruvian music draws on the country's heritage, taking percussion from Africa, wind instruments from the Amazon and string instruments (plus classical dance rhythms) from Spain. Los Mirlos are a fine example, drawing Afro-Caribbean *cumbia* and dreamy harmonics into a hip-twitching accessible sound.

Playlist

Te Mentiría
Gian Marco
Genre: Latin pop

Simiolo
Dengue Dengue Dengue
Genre: Electronica

Samba Malató
Lucila Campos
Genre: Criolla

Negra Presuntuosa
Susana Baca
Genre: Landó

La Danza de los Mirlos
Los Mirlos
Genre: Cumbia

Este Amor No Es Para Cobardes
Barrio Calavera
Genre: Ska

Se Acabó y Punto
Arturo 'Zambo' Cavero
Genre: Criolla

Quiero Amanecer
Bareto
Genre: Cumbia

Mal Paso
Eva Ayllón
Genre: Landó

Camino al Carnaval
Lucho Quequezana
Genre: Folk

Suriname

Read List

We Slaves of Suriname
by Anton de Kom (1934, trans 2016)
This anti-colonialist's critique of the cruelty of slavery made de Kom, the son of a slave, one of the Surinamese's most courageous advocates for social justice and dignity.

The Cost of Sugar
by Cynthia McLeod (1987, trans 2011)
The celebrated Surinamese novelist's exposé of life on an 18th-century sugar plantation, where the lives of two pampered Dutch step-sisters become entwined with that of their slaves (see the Dutch film, right).

The Free Negress Elizabeth
By Cynthia McLeod (2008)
In this novel, Elisabeth Samson, a free black Surinamese woman, overcomes institutionalised discrimination and prejudice to become one of the wealthiest individuals in the slave colony of 18th-century Dutch Guiana.

Wild Coast: Travels on South America's Untamed Edge
by John Gimlette (2011)
The English author's travelogue of his journey through the Guianas, which sees him venture deep into the jungles of Suriname, is guaranteed adventure inspiration.

Rainforest Warriors: Human Rights on Trial
by Richard Price (2011)
US anthropologist Price tells the gripping story of how Saramakas (one of six Maroon peoples of Suriname) harnessed international human rights law to win control of their own piece of the Amazonian forest.

Watch List

Wiren
(2018; drama; directed by Ivan Tai-Apin)
The first Surinamese film to be entered for an Academy Award, *Wiren* follows the story of a deaf boy who becomes a catalyst for change in a society where discrimination against people with disabilities runs deep.

One People
(1976; drama; directed by Pim de la Parra)
Centred on the relationship between an Afro-Surinamese man and a Hindu nurse, this iconic film explores the tense race relations in post-colonial Suriname.

The Cost of Sugar
(2013; drama; directed by Jean van de Velde)
The major motion picture adaptation of Surinamese novelist Cynthia McLeod's book – controversially shot in South Africa – offers one of the few cinematic windows into the dark days of Dutch colonial rule.

Stones Have Laws
(2019; documentary; directed by Tolin Alexander, Siebren de Hann & Lonnie van Brummelen)
This immersive collaboration with Suriname's Maroon community explores the legacy of slavery, and the folk stories attached to lives lived to the rhythm of the jungle.

Hidden World
(2018; documentary; directed by Kenrich Cairo)
In a remote Maroon village, Father Amoksi wants to transfer his gift of communicating with the spirit world onto his son – who is more interested in his phone.

Boats are often the only way to travel deeper into South America's wild interior.

Considered by many to be the mother of modern-day Chutney (Indo-Caribbean) music, Dropati's 1968 album *Let's Sing and Dance* has provided the soundtrack to many a sultry Hindu Caribbean wedding.

Known by the stage name Damaru, Surinamese rapper and singer Dino Orpheo Canterburg's collaboration with Dutch singer Jan Smit, *Mi Rowsu (My Rose)*, was a big hit in the Netherlands.

South America

Playlist

Raat Ke Sapna
Ramdew Chaitoe
Genre: Chutney

Pyaar Huwa
Freestyle
Genre: Hindi pop

Dui Mutthi
Raj Mohan
Genre: Pop

Gowri Pooja
Dropati
Genre: Folk/World

Na Foe Sang Èdè
Lieve Hugo
Genre: Kaseko

Vrede
Ruth Jacott
Genre: Pop

Mi Rowsu
Damaru
Genre: Reggae

Soul With Milk
Sumy
Genre: Disco

Parijs
Kenny B
Genre: Pop

Merengue Mix
Amit Sewgolam
Genre: Merengue/Baithak gana

Uruguay

Read List

The Open Veins of Latin America
by Eduardo Galeano (1971, trans 1973)
An enduringly popular economic history book that is broad in scope but zooms in on specific events, creating a vivid impression of how Latin America has been exploited.

The Truce
by Mario Benedetti (1960, trans 1969)
A series of diary entries written from the perspective of a 49 year-old widower whose life in Montevideo is transformed when, unexpectedly, he falls in love.

The Shipyard
by Juan Carlos Onetti (1961, trans 1968)
This widely celebrated but downbeat tale of an antihero's futile attempt to regenerate an abandoned shipyard has been interpreted as a comment on the decay of urban life and the breakdown of Uruguayan society.

The Decapitated Chicken and Other Stories
by Horácio Quiroga (1984)
Real-life experiences form the backdrop to many of Quiroga's often dark and rather bleak short stories, many of which were influenced by his time spent living in the jungle in the Argentine province of Misiones.

Piano Stories
by Felisberto Hernández (1993)
Piano playing features heavily and objects come to life in this anthology of short stories by a highly original writer, whose use of magical realism influenced the work of Gabriel García Márquez and Julio Cortázar.

Watch List

Whisky
(2004; drama; directed by Juan Pablo Rebella & Pablo Stoll)
With his brother due to visit, a lonely sock factory owner asks his employee to pose as his wife in this critically acclaimed, deadpan comedy.

Seawards Journey
(2003; comedy; directed by Guillermo Casanova)
Taking an affectionate, humorous and nostalgic look at rural Uruguay, this is the charming story of friends who take a day trip to the coast.

Anina
(2013; animation; directed by Alfredo Soderguit)
Ten-year-old Anina is given a strange punishment in an animation that perfectly captures the visual idiosyncrasies of everyday life for a child in Uruguay.

A Moonless Night
(2015; drama; directed by Germán Tejeira)
New Year's Eve in a small town in Uruguay is the setting for three stories connected by the themes of loneliness and the possibility of change.

A Twelve-Year Night
(2018; drama; directed by Álvaro Brechner)
A nuanced portrayal of the 12 years of solitary confinement endured by members of the Tupamaros revolutionaries – including José Mujica, who later became president of Uruguay – during the dictatorship of 1973 to 1985.

Friends set off for the remote Uruguayan coast in the film *Seawards Journey*.

The distinct rhythms and drum beats of *candombe* were first developed in the 19th century by the Afro-Uruguayan descendants of enslaved Africans. They combined African percussion with European and Caribbean instruments and rhythms to create a new musical form.

Bajofondo is a collaborative of eight musicians from Uruguay and Argentina that has achieved global success with its contemporary interpretations of traditional forms of music originating from the Río de la Plata region – not only tango, but also *candombe*, *murga* and *milonga*.

Playlist

Colombina
Jaime Roos
Genre: Murga

La cumparsita
Julio Sosa
Genre: Tango

Sea
Jorge Drexler
Genre: Pop

Muy Lejos Te Vas
El Kinto
Genre: Rock

Candombe para Gardel
Rubén Rada
Genre: Candombe

Pa' Bailar
Bajofondo
Genre: Neo-tango

Pasos
Malena Muyala
Genre: Pop

A las Nueve
No Te Va Gustar
Genre: Rock

Tu Vestido
Ana Prado
Genre: Folk

Doña Soledad
Alfredo Zitarrosa
Genre: Candombe

NO AME

RTH RICA

Barbados

Read List

In the Castle of My Skin
by George Lamming (1953)
Written in lyrical prose, this imaginative autobiographical novel describes nine year-old G coming of age in a fishing village as colonial rule begins to crumble around his ears.

The Star Side of Bird Hill
by Naomi Jackson (2015)
A pacy family drama set in the summer of '89 – two sisters are sent from their home in Brooklyn to live in Barbados with their grandmother, who is a practitioner of the local spirituality obeah.

Sugar in the Blood
by Andrea Stuart (2012)
Born in Barbados and the descendant of both slaves and slavers, Stuart uses her lineage as a lens with which to explore the impact of empire in this absorbing memoir.

Tracing JaJa
by Anthony Kellman (2016)
Full of evocative sense of place, this historical novel explores the relationship between an exiled African king and his Bajan servant, and is as much a love letter to the island as it is love story.

Strange Fruit
by Kamau Brathwaite (2016)
With a title referencing Billie Holiday's song protesting the lynching of black Americans, this collection by Barbados' most celebrated poet is a tribute to the creation of beauty in the face of adversity.

Watch List

Island in the Sun
(1957; drama; directed by Robert Rossen)
Shot in Barbados and chronicling a fictional island's uneasy relationship with their British colonisers, this film was controversial at the time for its interracial love story, depicted by stars Joan Fontaine and Harry Belafonte.

A Caribbean Dream
(2017; romantic comedy; directed by Shakirah Bourne)
Shakespeare set at carnival: Bourne's retelling of *A Midsummer Night's Dream* weaves Bajan folklore and cadence into the original text and features a predominantly local cast.

Barrow: Freedom Fighter
(2016; docudrama; directed by Marcia Weekes)
This impassioned biopic tells the story of Errol Barrow, an RAF airman-turned-lawyer who led Barbados to independence after three centuries as a British colony.

Sweet Bottom
(2016; drama; directed by Gladstone Yearwood)
Exploring the consequences of the US practice of deporting law-breaking Caribbeans, this micro-budget film follows a laconic army veteran as he tries to make a life in the country he left as a baby.

Hit for Six
(2007; sports; directed by Alison Saunders-Franklyn)
Giving insight into Barbados' beloved national sport, a sidelined West Indies cricket player battles his demons to win a place on the team and the respect of his father.

North America

Playlist

Tempted to Touch
Rupee
Genre: Reggaeton

Faluma
Alison Hinds
Genre: Soca

Walk Away from Love
Shirley Stewart
Genre: Soul

Drink Milk
The Draytons Two
Genre: Spouge

Pon de Replay
Rihanna
Genre: Pop

One of the best-selling musical artists of all time, Robyn Rihanna Fenty released her debut single in 2005. *Pon de Replay* means 'play it again' in Bajan Creole, and reflects her early love of popular local styles such as dancehall and reggae.

Jack
Mighty Gabby
Genre: Calypso

Cry Me a River
Jackie Opel
Genre: Spouge

Released in 1965, *Cry Me a River* is a beautifully sung ballad and one of the bigger hits by musical pioneer Jackie Opel. Credited with inventing 'spouge' music, he blended Trinidadian calypso with Jamaican ska to create a uniquely Bajan sound.

Mr. Harding
Red Plastic Bag
Genre: Calypso

Who the Hell is Kim?
Terencia Coward
Genre: Soca

Twilight
Cover Drive
Genre: Pop

A local cricket match in Barbados; cricket is a Caribbean obsession, as depicted in *Hit for Six*.

Canada

Read List

The Blind Assassin
by Margaret Atwood (2000)
A complex novel within a novel within a novel that uses plot twists and time shifts to frame a story of romance and mystery. The backdrop is real-life historical events.

Lives of Girls and Women
by Alice Munro (1971)
Nobel Prize-winner Munro is a master of interconnected short stories and these eight tales all focus on the same character – Del Jordan coming of age in a small town in Southern Ontario.

Anne of Green Gables
by Lucy Maud Montgomery (1908)
A cornerstone of Canadian literature and a massive tourist lure for Prince Edward Island, the story of orphan Anne Shirley and her adventures at Avonlea farm has sold more than 50 million copies.

In the Skin of a Lion
by Michael Ondaatje (1987)
The celebrated Canadian-Sri Lankan author shines a light on Canadian immigrants and their role in building Toronto during the early 1900s. The book's sequel, *The English Patient*, would go on to win the Booker Prize in 1992.

Fifth Business
by Robertson Davies (1970)
The first book in Davies' *Deptford Trilogy* is set in a fictional Ontario village and relates the impact of a misplaced snowball that sets off a chain of events.

Watch List

Mon Oncle Antoine
(1971; drama; directed by Claude Jutra)
There's a moody French arthouse aesthetic to this slow-moving, ultra-realistic study of life in a snowy mining town in rural Quebec, as seen through the eyes of a 15-year-old boy.

Goin' Down the Road
(1970; drama; directed by Donald Shebib)
Low-budget hippy-era flick about two working-class dudes from Nova Scotia who come to Toronto looking for riches and excitement but end up broke and in trouble.

Dead Ringers
(1988; thriller; directed by David Cronenberg)
Set in Cronenberg's native city, Toronto, Jeremy Irons does double duty playing gynaecologist twins in a creepy tale of impersonation, drug abuse and weird delusions that doesn't end well for either of them.

Jesus of Montreal
(1989; comedy drama; directed by Denys Arcand)
A biblical allegory about a struggling Montreal actor trying to put on a modern Passion Play with complicated and often comic results. The French-Canadian classic won a Jury Prize at Cannes.

The Sweet Hereafter
(1997; drama; directed by Atom Egoyan)
A school bus crash in rural British Columbia leads to lawsuits, witness tampering and the unmasking of bitter small-town secrets.

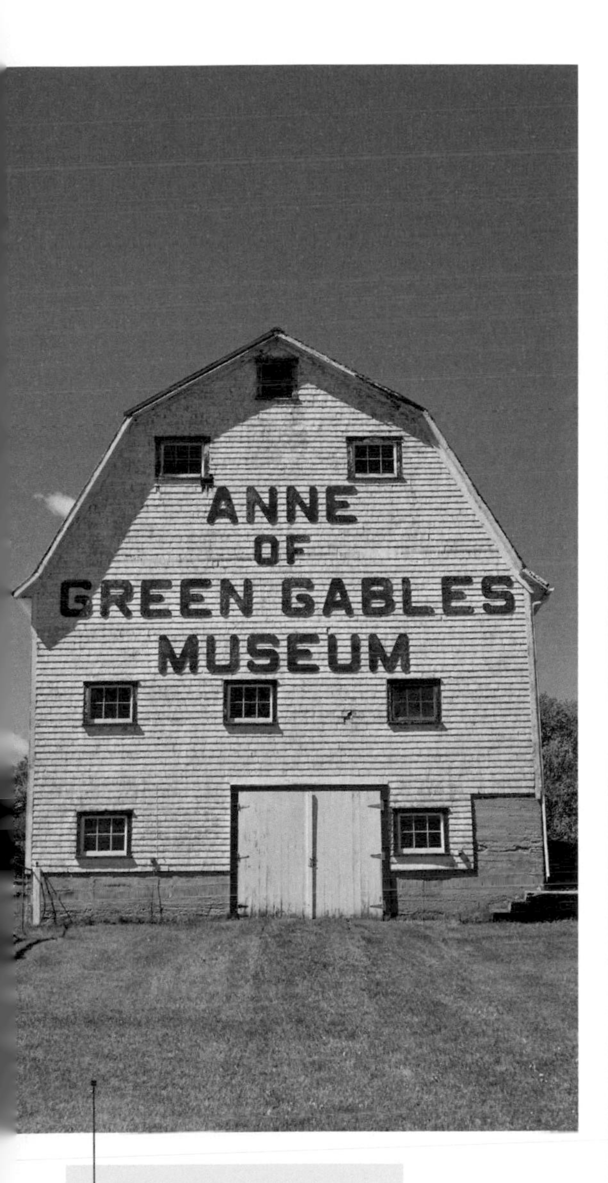

On Prince Edward Island you're in charming *Anne of Green Gables* country.

Wildly popular in Canada but little known elsewhere, The Tragically Hip are a national institution whose poetic lyrics – written by their late singer Gord Downie – are loaded with wit, pathos, honesty and astute cultural references summing up what it means to be Canadian.

While the US had Bob Dylan, Canada countered with Cohen, a Fedora-wearing street poet and social commentator from Montreal known for his gravelly voice and ability to look cool well into his '70s. His most famous song, *Hallelujah*, has been recorded by over 300 different artists.

Playlist

Both Sides Now
Joni Mitchell
Genre: Folk rock

The Suburbs
Arcade Fire
Genre: Rock

The Weight
The Band
Genre: Roots rock

The Needle and the Damage Done
Neil Young
Genre: Folk rock

Ahead by a Century
The Tragically Hip
Genre: Rock

Hallelujah
Leonard Cohen
Genre: Folk rock

Blinding Lights
The Weeknd
Genre: Electropop

You Oughta Know
Alanis Morissette
Genre: Alt rock

God's Plan
Drake
Genre: Rap

Hymn to Freedom
Oscar Peterson
Genre: Jazz

Cayman Islands

Read List

Caribbean Cartels
by Sheldon M Brown (2010)
The first in a series of crime thrillers written by a reformed local gangster, Brown brings authenticity to tales set in the violent underworld inhabited by Cayman drug-traffickers.

Reflections from a Broken Mirror: Poems about Caymanian Society
by J A Roy Bodden (2014)
The author of several books on Caymanian history, Bodden turns to poetry to express his anguish at the exploitation of his beloved islands by foreign powers.

The Confessions of Frannie Langton
by Sara Collins (2019)
Collins, who grew up in Grand Cayman, explores themes of race, gender and the psychology of servitude in her acclaimed murder mystery, set partially in the Caribbean.

Second Wind
by Dick Francis (1999)
The British jockey-turned-crime writer saw out his last years in Cayman, where he partly sets this fanciful tale of a meteorologist who discovers a dark secret when his plane crash-lands on a Caribbean island.

Outposts
by Simon Winchester (1985)
This engaging and often funny travelogue includes the British-American journalist's reflections on visiting Grand Cayman as he explores relics of the British Empire.

Watch List

Haven
(2004; thriller; directed by Frank E Flowers)
The Cayman Islands' biggest cinematic success, made by its most celebrated director, stars Hollywood names including Orlando Bloom and Zoë Saldaña.

Cayman Went
(2009; romance; directed by Bobby Sheehan)
Full of alluring coastal scenes and impressive underwater photography, this indie film centres on a fading action star learning important life lessons among the eccentric diving community of Cayman Brac.

The Firm
(1993; thriller; directed by Sydney Pollack)
In this hit John Grisham adaptation starring Tom Cruise, scenes shot on Grand Cayman – including at Seven Mile Beach – expound on the islands' infamy as a tax haven.

HER Story is OUR History
(2017; documentary; directed by Sean Bodden)
An account of the Cayman Islands' women's suffrage movement between 1948 and 1958, with first-hand testimony and reflections from those involved.

Bright Spot
(2015; documentary; directed by Tamer Soliman & Rob Tyler)
A nutritionist explores why Caymanians have largely stopped using coconut oil, despite its traditional use in local cuisine and the health benefits that have led to it being marketed as a 'superfood' elsewhere.

The waters around the Cayman Islands are scattered with shipwrecks, including the USS *Kittiwake*.

Cotterell is the Caymans' most popular home-grown artist and has worked with American hip-hop star Flo Rida as a producer.

Swanky Kitchen Band is a multi-instrument group dedicated to preserving the Cayman Islands' tradition of 'kitchen dance' music – a genre that combines Celtic fiddle song with soulful African rhythms and calypso, reggae and jazz. In recognition of the local custom for Christmas Eve serenading, seasonal music is an important part of their repertoire.

Playlist

Hunnid
Josh Pearl
Genre: Rap

Soul surfer
Natasha Kozaily
Genre: Pop

Party Time
Cotterell ft Popeye Caution
Genre: Dancehall

Real Come Back Story
Stuart Wilson
Genre: Reggae

Munzie Boat in the Sound
Swanky Kitchen Band
Genre: Kitchen dance

Wicked Game
Hi Tide
Genre: Acoustic

Martyr
Maeve
Genre: Pop

Tigers
Suckerbox
Genre: Punk

Man Up
Andrea Rivera
Genre: Soca

Spice
Precious Shordee
Genre: Hip-hop

North America

Costa Rica

Read List

God Was Looking the Other Way
by José León Sánchez (1963, trans 1973)
Illiterate when he was sent off to a notorious prison, Sánchez taught himself how to read and write, and clandestinely authored one of Latin America's most compelling memoirs.

There Never Was a Once Upon a Time
by Carmen Naranjo (1984, trans 1989)
Written from a child's perspective, these first-person stories sparkle with insight, as youths find solace in imaginary realms while grappling with growing up.

The Best Short Stories of Quince Duncan
by Quince Duncan (1995)
Costa Rica's most prominent Afro-Caribbean writer paints a revealing portrait of customs and culture while dealing with racism and exclusion, in this beautifully written, dual-language collection of stories.

Monkeys Are Made of Chocolate
by Jack Ewing (2003)
Written by an American naturalist and long-time Costa Rica resident, these short, thought-provoking essays blend environmental science with rural culture in jungleclad Puntarenas province.

The Costa Rica Reader
edited by Iván Molina and Steven Palmer (2004)
This collection of over 50 essays, written mostly by Costa Ricans, brings the nation's culture and history to life, delving into culture, politics, the arts and nature.

Watch List

Of Love and Other Demons
(2009; drama; directed by Hilda Hidalgo)
Costa Rican film-maker Hidalgo adapts Gabriel García Márquez's magical-realism novel for the big screen, in this Costa Rican-Colombian production about a priest who falls in love with a condemned young teenager.

Maikol Yordan Traveling Lost
(2014; comedy; directed by Miguel Alejandro Gomez)
This much-loved Tico classic features a warm-hearted but naive farmer who travels abroad in hopes of raising money to save the family property.

A Bold Peace
(2016; documentary; directed by Michael Dreiling & Matthew Eddy)
This inspiring human rights film details Costa Rica's courageous decision (made in 1948) to abandon its military program, and invest instead in healthcare, education and the environment.

Viaje
(2015; drama; directed by Paz Fábrega)
Shot in black-and-white, this short film captures the playfulness and nostalgia of a new romance, much of which unfolds in lush Rincón de la Vieja National Park.

Caribe
(2004; drama; directed by Esteban Ramírez)
Costa Rica's pristine coastline stars in this saga of love and greed, starring a banana-farming couple and a corrupt oil company.

Squirrel monkeys are some of the wild creatures you'll encounter in Costa Rica's jungles.

An icon of calypso music, Walter Ferguson had a career spanning more than seven decades. His songs are full of humour, tragedy and humility – he lived most of his life in the fishing village of Cahuita, where he was raised.

Well-known Cantoamérica was formed in 1980 as part of 'the new Costa Rican song' movement that celebrated music with Latin American roots. Taking the message to heart, Cantoamérica fuses a wide range of musical traditions, including calypso, rumba, bolero, reggae and other Caribbean rhythms.

Playlist

The Feel
Las Robertas
Genre: Garage rock

No Nos Sobran los Domingos
Debi Nova
Genre: Pop

El Invisible
Gandhi
Genre: Rock

Cabin in the Water
Walter Ferguson
Genre: Calypso

La Cura
Los Ajenos
Genre: Pop/Rock

Afrolimon
Cantoamérica
Genre: Caribbean fusion

Terremoto
Passiflora
Genre: Gypsy-folk

Chúcaro
Cocofunka
Genre: Rock

Ya No
Ojo de Buey
Genre: Reggae

Supernova
Alphabetics
Genre: Indie rock

Cuba

Read List

Cecilia Valdés
by Cirilo Villaverde de la Paz (1839, trans 2005)
Considered the first Cuban novel, *Cecilia Valdés* recounts the ill-fated love between a slave trader's illegitimate black daughter and the man's legitimate son. It is a brave look at the complexities of race and class in 19th-century Cuba, only published in full in New York in 1882.

Simple Verses
by José Martí (1891, trans 1996)
A revolutionary instrumental in liberating Cuba from Spanish rule, Martí was also a poet. This elegiac collection draws on themes pertinent in the Cuba of that era.

Explosion in a Cathedral
by Alejo Carpentier (1962, trans 1963)
Outside a Havana mansion, a storm of revolution is breaking across the Caribbean. Action swings rivetingly from Cuba to Guadeloupe to French Guiana.

Before Night Falls
by Reinaldo Arenas (1992, trans 1993)
A young gay writer from Eastern Cuba tries to find himself in 1970s Havana: cue a blackly humorous string of relationships, brushes with the law and much more.

Trilogía sucia de La Habana (Dirty Havana Trilogy)
by Pedro Juan Gutiérrez (1998, trans 2002)
Gutiérrez is a key exponent of 'dirty realism', as this sultry trilogy reveals. It details the protagonist's liaisons, depression and grim depictions of life in Cuba's capital during the economic depression of the 1990s.

Watch List

I am Cuba
(1964; historical drama; directed by Mikhail Kalatozov)
This moody black-and-white montage of Cuban responses to oppression was a Soviet-Cuban venture. It was initially dismissed by Soviet audiences but later praised for its filming techniques.

Death of a Bureaucrat
(1966; comedy; directed by Tomás Gutiérrez Alea)
This playful look at the ridiculousness of Communist bureaucracy still resonates in Cuba; it's regarded among the greats of Cuban movie-making.

Memories of Underdevelopment
1968; drama; directed by Tomás Gutiérrez Alea)
Narrator Sergio Corrieri, an upper-class intellectual, remains in Cuba although his wife and friends have left for Miami. This stylistic film follows his reminiscences, Cuban Missile Crisis to personal relationships.

Lucía
(1968; historical drama; directed by Humberto Solás)
Three different women, each named Lucía, relate three different chapters in Cuban history, each superbly acted.

Strawberry and Chocolate
(1993; comedy drama; directed by Tomás Gutiérrez Alea & Juan Carlos Tabío)
Havana 1979: gay artist Diego unsuccessfully attempts seducing straight, prim David, who then agrees to cultivate a friendship anyway to monitor Diego for the state. An Oscar-nominated heartstring-puller.

Havana seems to teeter on the cusp between beauty and decay and has inspired numerous writers.

The general consensus is that Trío Matamoros are Cuba's most seminal *son* band of all time, and that *Lagrimas Negras* is THE perfect fusion of *son* and *bolero*.

The original version of *Chan Chan*, the most-played Cuban song ever, purportedly came to veteran trova singer-songwriter-musician Compay Segundo in a dream in 1984. However it was the Buena Vista Social Club – an ensemble established in 1996 to revive the music of pre-Revolutionary 'golden age' Cuba (from the 1930s to '50s) – that recorded the mood-setting, instantly recognisable classic version of *Chan Chan*.

Playlist

Guantanemera
Cuarteto Caney
Genre: Son

Contradanza
Chucho Valdés
Genre: Jazz

Lagrimas Negras
Trío Matamoros
Genre: Son

Santa Isabel de las Lajas
Benny Moré
Genre: Son

Hasta Siempre
Carlos Puebla
Genre: Trova

La Vida es un Carnaval
Celia Cruz
Genre: Salsa

El Baile del Buey Cansao
Los Van Van
Genre: Songo

Muros y Puertas
Carlos Varela
Genre: Nueva trova

Chan Chan
Compay Segundo & Buena Vista Social Club
Genre: Son

Havana
Camila Cabello ft Young Thug
Genre: R&B/Latin pop

IN DEPTH

Cuban Son

from CUBA

If Cuban music had a backbone, *son Cubano* would be it. Hailing from the Eastern Cuba (*oriente*) countryside towards the end of the 19th century, *son* developed in a climate where *danzón* was the queen of Cuban dances. *Danzón* was itself the Spanish-American take on the contradance, which had originated in English country dancing to become by the 18th century the world's most recognised dance style. With its catchier clave rhythmic arrangement and use of Africa-originating instruments like the bongo, *quijada* (percussion instruments originally made from donkey jawbone) and *marimbula* (box-like string instruments), *son* was synonymous with *oriente* plantations and the enslaved Africans who worked there:

when slavery was abolished many plantation workers moved to Havana where *son* supplanted *danzón* as Cuba's most popular music. By the 1920s it was typically played in a sextet featuring guitar, *tres* (a guitar-like instrument with three pairs of strings) claves, maracas, bongos and double-bass. Titanic early exponents were Trío Matamoros who partly synthesised instruments found in traditional sextets and adapted the genre to give it a diverse sound range, infusing styles like bolero. *Son* virtuoso 'Bárbaro del Ritmo' (Barbarian of Rhythm!) Benny Moré later mixed in mambo and *guaracha* to the style. Pure *son* waned after Trío Matamoros' passing but the legacy survives in jazz, salsa and myriad other types of Cuban music.

Left to right: a Cuban band playing in a Havana street; dancing to *danzón*; acclaimed musician Compay Segundo and the Buena Vista Social Club band, influenced by Cuban *son*.

Guatemala

Read List

Men of Maize
by Miguel Ángel Asturias (1949, trans 1975)
The Nobel Prize-winning author's complicated masterpiece is a six-part epic that delves into the mystical world of an indigenous community and its fight for survival against foreign invaders.

A Mayan Life
by Gaspar Pedro González (1995)
The first ever novel published by a Q'anjob'al speaker gives deep insight into the cultural traditions in an indigenous highland village.

Popol Vuh
author unknown (pre-1500)
Recounting the creation myths and legendary ancestors of the Quiché people, this astonishing 'Mayan Bible' predates the Spanish conquest and was originally preserved through oral traditions.

Complete Works and Other Stories
by Augusto Monterroso (1995)
Guatemala's literary master blends surrealism with sardonic wit. This collection brings together stories first published in 1959 and 1972, roaming from philosophy to whimsy and twisted anthropological tales.

The Long Night of White Chickens
by Francisco Goldman (1992)
The tragic story of a Guatemalan orphan who creates a new life in the US and later returns to Guatemala brilliantly captures the dark, violent times of the 1980s.

Watch List

Ixcanul
(2015; drama; directed by Jayro Bustamante)
The first film ever made in the Kaqchikel Mayan language is a luminous, hypnotic work of one woman torn between tradition and modernity.

La Camioneta
(2012; documentary; directed by Mark Kendall)
A brilliant, ruminating film that follows the transformative journey of a decommissioned US school bus to its rebirth as public transport in Guatemala.

When the Mountains Tremble
(1983; documentary; directed by Pamela Yates & Newton Thomas Sigel)
This powerful work describes the massacres of indigenous communities by the Guatemalan military. It features Nobel Peace Prize-winner Rigoberta Menchú, a Quiché human-rights activist.

José
(2018; drama; directed by Li Cheng)
A gripping, neorealistic drama about the struggles of one young gay man coming of age amid the machismo and church dominance of Guatemala City.

September
(2017; drama; directed by Kenneth Muller)
A film that gives a face to the victims of Guatemala's armed conflict, this affecting coming-of-age story features a girl left deaf and motherless following a terrorist attack.

Guatemala's indigenous Mayan culture can be felt at ruins such as Tikal.

The song *Guatemaya* celebrates indigenous culture with its haunting *marimba* rhythms (a favourite in Mayan celebrations), references to 'burning copal' (used in spiritual ceremonies), and the call to end genocide against native people. The track also features rapper Tz'utu Kan, who rhymes in the ancient Mayan language of Tz'utujil.

Pioneering Guatemalan artist Sara Curruchich sings both in the indigenous Mayan language of Kaqchikel (her mother tongue) and in Spanish. An activist for human rights, she has performed around the world; in Guatemala she serves as an inspiration to indigenous communities.

Playlist

Blues de Mar
Gaby Moreno
Genre: Blues

Aire
Bohemia Suburbana
Genre: Alt rock

Ni Una Menos
Rebeca Lane
Genre: Hip-hop

Animal Nocturno
Ricardo Arjona
Genre: Pop

Guatemaya
Doctor Nativo
Genre: Latin fusion

Mismos Ojos
Easy Easy
Genre: Indie rock

Somos
Sara Curruchich
Genre: Traditional

Luna Llena
Malacates Trebol Shop
Genre: Pop/Rock

Yunduha Weyu
Sofia Blanco
Genre: Traditional

Luna de Xelaju
Paco Pérez
Genre: Waltz

Haiti

Read List

Breath, Eyes, Memory
by Edwidge Danticat (1994)
Catapulting Danticat to fame when it was selected for Oprah's Book Club, the debut novel of Haiti's most celebrated writer is a heartbreaking tribute to the strength of her countrywomen.

Hadriana in All My Dreams
by René Depestre (1988, trans 2017)
Set during carnival in the Haitian town of Jacmel in 1938, this novel's action begins with its heroine's death – and her resurrection. Gloriously odd, Depestre's vodou love story earned him the prestigious Prix Renaudot.

Love, Anger, Madness
by Marie Vieux-Chauvet (1968, trans 2010)
Suppressed when it was first published, Vieux-Chauvet's trilogy of novellas were so critical of life under 'Papa Doc' Duvalier's repressive rule she was forced to flee to the US.

The Enigma of the Return
by Dany Laferrière (2013)
Combining fiction, poetry and autobiography, this work follows an exiled writer's return to Haiti after 30 years. Capturing both the country's poverty and its enduring beauty, it's also a moving exploration of loss and identity.

Nan Domi: An Initiate's Journey into Haitian Vodou
by Mimerose Beaubrun (2013)
This anthropological writer's account of her initiation into and education in vodou is full of revelatory observations on a misunderstood religion.

Watch List

Fatal Assistance
(2013; documentary; director Raoul Peck)
Filmed during the two years following Haiti's devastating 2010 earthquake, the country's former Minister for Culture examines the further damage inflicted by international aid agencies.

Stones in the Sun
(2012; drama; director Patricia Benoit)
Following the interlocking stories of three different families fleeing political violence, Benoit's powerful film uses a cast of mostly untrained actors.

Heading South
(2005; drama; director Laurent Cantet)
Based on a novel of the same name by Dany Laferrière, *Heading South* stars Charlotte Rampling as a middle-aged sex tourist visiting Haiti during the toxic political climate of the late 1970s.

Jacques Roumain: Passion for a Country
(2008; docudrama; directed by Antonin Arnold).
One of Haiti's most prolific filmmakers explores the life of a Haitian intellectual heroe – Roumain created an enduring canon of politically conscious novels and poetry before his early death, aged 37, in 1944.

Cousines
(2006; drama; directed by Richard Sénécal).
Following the fortunes of a young girl in Port-au-Prince who is sent to live with a friend after her father dies, this film explores the personal impact of political instability.

Wyclef Jean has transitioned from musician to popular leader and campaigner in Haiti.

Haiti's biggest musical export, Wyclef Jean released his debut solo album *The Carnival* in 1997, after finding success with The Fugees. This is one of several tracks performed in Haïtian Creole, and features former bandmate Lauryn Hill

Grammy-nominated for their first album, the 'Haïtian roots music' band Boukman Eksperyans are named after two of their key influences – 18th-century vodou priest Dutty Boukman and legendary guitarist Jimi Hendrix.

Playlist

Nostalgie Haïtienne
Martha Jean-Claude
Genre: Folk/Calypso

Haïti Chérie
Haïtian Troubadours
Genre: Twoubadou

Haïti Cumbia
Nemours Jean-Baptiste
Genre: Compas

Yele
Wyclef Jean
Genre: Hip-hop

Tu as volé
Tabou Combo
Genre: Mini-jazz/Compas

Jwi Lavi
Barikad Crew
Genre: Rap Kreyol

Ke'm Pa Sote
Boukman Eksperyans
Genre: Mizik rasin

Timoun
Emeline Michel
Genre: Folk

Akwaaba Ayiti
Michael Brun
Genre: Dance

Ibo lé lé
Jephté Guillaume
Genre: Electronic

Jamaica

Read List

A Brief History of Seven Killings
by Marlon James (2014)
The first Jamaican novel to win the Booker Prize explores the attempted assassination of Bob Marley in 1976 in complex detail, employing a huge cast of characters both fictional and real-life.

White Witch of Rose Hall
by H G de Lisser (1929)
A gothic horror story based on a popular legend about the spirit of Annie Palmer, who supposedly haunts the old plantation house of Rose Hall where she murdered three of her husbands.

Here Comes the Sun
by Nicole Dennis-Benn (2017)
The other side of paradise: this novel is told through the eyes of an employee at an all-inclusive Jamaican resort as she battles poverty, homophobia and misogyny.

Waiting in Vain
by Colin Channer (1998)
Named after a Bob Marley song, this engrossing tale speckled with colourful patois follows two protagonists as they journey through Kingston and the Jamaican diaspora in London and New York.

The Lunatic
by Anthony Winkler (1987)
A comic novel revolving around a village madman who has an affair with a tourist. Winkler's second book became a bestseller and was subsequently made into a 1991 film.

Watch List

The Harder They Come
(1972; crime; directed by Perry Henzell)
Jamaica's first feature film was a sharp, rugged crime thriller that made a star of Jimmy Cliff and put reggae music on the cultural map.

Marley
(2012; documentary; directed by Kevin Macdonald)
Scottish director Macdonald assembled an exemplary collection of contemporary interviewees, never-seen-before concert footage and fabulous music for this inspiring biopic of Jamaica's greatest musician.

Better Mus' Come
(2010; drama; directed by Storm Saulter)
The horrific gang warfare of 1970s Jamaica provides a dramatic backdrop for Saulter's feature film debut that adds a message of hope with a love story and some artistic cinematography.

Smile Orange
(1976; comedy; directed by Trevor Rhone)
This satirical movie based on a stage play follows the escapades of Jamaican con man Ringo, a waiter in a tourist hotel, as he tries to manipulate his exploiters.

Third World Cop
(1999; action, directed by Chris Browne)
Produced by Island Records founder Chris Blackwell, this high-grossing crime flick follows a Jamaican cop as he investigates an arms trafficking ring run by one of his childhood friends.

Bob Marley is the best-known exponent of Jamaica's rich musical outpouring.

Marley's first big international hit made the top 10 in the UK and became a reggae classic. With his gritty descriptions of eating cornmeal porridge in Kingston's Trenchtown, Marley took Jamaican music to the world while raising awareness about Rastafarianism, Pan-Africanism and social justice.

Wayne Smith's revolutionary form of dancehall music mixed elements of reggae with an electronic 'riddim' borrowed from a Casio keyboard. The hypnotic sound, known in the UK as 'ragga', had many imitators, but few were as good as *Sleng Teng* – Smith's ode to marijuana.

Playlist

Many Rivers to Cross
Jimmy Cliff
Genre: Reggae

007 Shanty Town
Desmond Dekker
Genre: Rocksteady

Rudy, A Message to You
Dandy Livingstone
Genre: Rocksteady

What is Life?
Black Uhuru
Genre: Reggae

No Woman, No Cry
Bob Marley & the Wailers
Genre: Reggae

Under Me Sleng Teng
Wayne Smith
Genre: Dancehall

Revolution
Dennis Brown
Genre: Reggae

King Tubby Meets Rockers Uptown
Augustus Pablo
Genre: Dub

Murder She Wrote
Chaka Demus & Pliers
Genre: Dancehall

Cheerleader
OMI
Genre: Reggae fusion

Mexico

Read List

The Labyrinth of Solitude
by Octavio Paz (1950, trans 1961)
Mexican poet, diplomat and Nobel Prize-winner Paz's essays explore the country's history and politics, and argue that solitude is at the heart of its identity.

Pedro Páramo
by Juan Rulfo (1955)
Many argue that magic realism began here, in a wild, poetic novel that tells of a man who travels to meet his mother after his father's death – and finds a perplexing ghost town.

The Death of Artemio Cruz
by Carlos Fuentes (1962, trans 1964)
A soldier-turned-politician looks back at a life of conflict and betrayal, in a novel inspired by *Citizen Kane* that helped spark a wave of interest in Latin American fiction.

Like Water for Chocolate
by Laura Esquivel (1989, trans 1993)
Recipes dot the pages of Esquivel's acclaimed fable of doomed love and great food, which was made into a hit film in 1992 and has sold over a million copies.

Loop
by Brenda Lozano (2014, trans 2019)
This novella, from one of the country's best young authors, takes the form of a Mexico City writer's journal. It covers everything from gang violence to literary classics in an innovative meditation on life and creativity.

Watch List

You're Missing the Point
(1940; comedy; directed by Juan Bustillo Oro)
A mistaken-identity classic, and the film that made Cantinflas (often described as Mexico's Chaplin) a star.

The Young and the Damned
(1950; drama; directed by Luis Buñuel)
Social realism meets surreal fantasy in this account of street kids struggling to survive in Mexico City. It's an unflinching portrayal of a dog-eat-dog world.

Cronos
(1993; horror; directed by Guillermo del Toro)
Before Hollywood came calling, del Toro directed this elegantly gruesome take on vampirism, featuring an ancient artefact, an antique dealer and his granddaughter.

Amores Perros
(2000; drama; directed by Alejandro González Iñárritu)
This stylish, violent and thought-provoking triptych follows dogfights, a troubled relationship and a hitman, before bringing the three together with a bang.

Roma
(2018; drama; directed by Alfonso Cuarón)
Cuarón's domestic drama has a wealthy Mexico City family and their maid at its heart; it's a soulful and stunning piece of film-making that deservedly won more than one Oscar.

Laura Esquivel, author of *Like Water for Chocolate*, was born in Mexico City and has moved into politics.

La Bamba is a traditional Veracruz folk song that has seen many interpretations over the years. US singer Ritchie Valens fused his Mexican heritage with amped-up rock'n'roll for this 1958 version, itself later covered by Los Lobos.

Thalía is Mexican royalty – as 'Queen of Latin Pop' she's sold 25 million records, while as 'Queen of Telenovelas' she's the genre's best paid actress, with her dramas shown in more than 180 countries.

Playlist

Mexico Lindo y Querido
Jorge Negrete
Genre: Ranchera

Ni Una Sola Palabra
Paulina Rubio
Genre: Pop

Sí Señor
Control Machete
Genre: Hip-hop

La Bamba
Ritchie Valens
Genre: Rock 'n' roll

Black Magic Woman
Santana
Genre: Rock

Nada Personal
Armando Manzanero
Genre: Bolero

Sinfonia India
Carlos Chávez
Genre: Classical

No Me Acuerdo
Thalía
Genre: Pop

Cuando Calienta El Sol
Luis Miguel
Genre: Pop

Pachuco
Maldita Vecindad
Genre: Ska

IN DEPTH

New Mexican Cinema

from MEXICO

© ULRIKESTEIN / GETTY IMAGES

Left to right: the Roma Sur neighbourhood of Mexico City; Penelope Cruz and Goya Toledo, stars of *Amores Perros*; directors Alfonso Cuarón, Alejandro González Iñárritu and Guillermo del Toro.

© RON GALELLA / GETTY IMAGES

Mexican cinema's first golden age, from the 1930s to 1960s, saw the country lead Latin America and capture a sizable audience around the world. But, as TV grew in popularity and American and French filmmakers grew increasingly successful, Mexico's crown slipped. The low-budget 'Mexploitation' films of the decades that followed had their highlights – including Robert Rodriguez's lean, violent *El Mariachi* (1992) – but it was a rare global hit.

Then, in the '90s and '00s, a brilliant new generation rose. First came Guillermo del Toro's cult horror *Cronos*, and a decade later Alejandro González Iñárritu's gritty *Amores Perros* and Alfonso Cuarón's joyous *Y tu Mamá También* (2001) were instant classics. Hollywood took notice: del Toro moved on to *Pan's Labyrinth* (2006; three Oscars) Cuarón to *Gravity* (2012; seven Oscars), and Iñárritu to *Birdman* (2014; four Oscars).

Homegrown films have continued to soar. Pedro González-Rubio's *Alamar* (*To the Sea*; 2009) is a gentle exploration of father-son, while Michel Franco's *Después de Lucía* (*After Lucia*; 2012) is an uncomfortable look at high-school bullying. In 2017, Ernesto Contreras' *Sueño en otro idioma* (*I Dream in Another Language*) pondered the demise of Indigenous languages. And local stories continued to be told by the biggest names: Cuarón returned to Mexico City with *Roma*, which took the 2018 Best Picture Oscar.

Nicaragua

Read List

The Jaguar Smile: A Nicaraguan Journey
by Salman Rushdie (1987)
Rushdie visited Nicaragua in 1986, meeting politicians, farmers and soldiers, and penning this broadly sympathetic account of life under the Sandinista government.

The Country Under My Skin: A Memoir of Love and War
by Gioconda Belli (2000, trans 2002)
Belli had a cloistered upper-class Managua upbringing, before smuggling arms and raising funds for the Sandinistas' struggle against the Somoza dictatorship.

Wild Shore: Life and Death with Nicaragua's Last Shark Hunters
by Edward Marriott (2000)
The bull shark, which swims up the San Juan river to Lake Nicaragua, is a national emblem that's fallen on hard times. Marriott follows its trail in this travelogue, taking in myths, fishing boats and rusting factories.

Selected Writings
by Rubén Darío (2005)
Late-19th-century poet Darío is considered the father of Spanish modernism, and this collection combines graceful verse with essays, short stories and travelogue.

Pluriverse: New and Selected Poems
by Ernesto Cardenal (2009)
A priest who became Nicaragua's Minister of Culture, Cardenal's celebrated poetry mixes politics, romance and metaphysics. This collection spans six decades of his work.

Watch List

Alsino and the Condor
(1982; drama; directed by Miguel Littín)
The first Nicaraguan film to be widely distributed, this is a lyrical account of a boy who dreams of flying as war rages around him.

Nicaragua Was Our Home
(1985; documentary; directed by Lee Shapiro)
American film-maker Shapiro presents a rarely told side of Nicaragua's civil war: that of the Miskito people of the Atlantic Coast, who suffered brutal violence during the conflict.

Nicaragua: A Nation's Right to Survive
(1983; documentary; directed by Alan Lowery)
Australian writer John Pilger probes American interventions in Nicaragua, in a passionate indictment of US support for the Somoza dictatorship and Contra rebels.

La Yuma
(2009; drama; directed by Florence Jaugey)
A girl sees boxing as an escape from her grim *barrio* in the captial of Managua, but her progress is complicated when she falls in love, in this rare Nicaraguan-made feature.

Kill the Messenger
(2014; drama; directed by Michael Cuesta)
Jeremy Renner stars as an American reporter in this intelligent, angry exposé of the CIA's role in Contra rebels' cocaine smuggling.

Nicaragua's capital, Managua, is the tough setting for Florence Jaugey's film *La Yuma*.

Luis Enrique moved to the US aged 16, but has proud Nica roots. The 'prince of salsa' is the nephew of *nueva canción* stars Carlos and Luis Enrique Mejía Godoy, and his romantic take on the genre saw him win a Grammy for Best Tropical Latin Album in 2009.

Marimba, Latin pop and reggaeton are popular all over Nicaragua, but the country's Atlantic Coast has a more Caribbean feel. Here, reggae is hugely popular, as is the festive *palo de mayo* dance, which originated in the town of Bluefields and mixes mento folk music with calypso.

Playlist

Pochote
Clara Grun
Genre: Pop

Anclado Al Aire
Perrozompopo
Genre: Alt-Latin

Flor de Mi Colina
Camilo Zapata
Genre: Marimba

Yo No Sé Mañana
Luis Enrique
Genre: Salsa

Nicaragua, Nicaragüita
Carlos Mejía Godoy
Genre: Marimba

The Dump
Soul Vibrations
Genre: Reggae

Come Down Brother Willy
DJ Maya ft Dimension Costeña
Genre: Cumbia

Dale Una Luz
Duo Guardabarranco
Genre: Folk

Ruinas
José de la Cruz Mena
Genre: Classical

Guaguanco in Japan
José 'Chepito' Areas
Genre: Jazz fusion

Panama

Read List

The World in Half
by Cristina Henríquez (2009)
As a young woman travels from Chicago to Panama City in search of a father she never knew, Henríque creates a compelling portrait of Panama's people and landscapes.

The Path Between the Seas
by David McCullough (1977)
The acclaimed American author describes in vivid detail the people, places and events involved in the mammoth undertaking of building the Panama Canal. Extensive photographs accompany this richly told history.

The Prince
by R M Koster (1972)
A Panama resident since 1957, Koster writes of power and politics with an ample dose of magical realism in this sardonic tale set in the imaginary Central American republic of Tinieblas.

The Golden Horse
by Juan David Morgan (2014)
Panamanian scholar Morgan writes brilliantly of the triumph and tragedy involved in building the first transcontinental railway in the Americas, carved through dense jungle during the California Gold Rush.

Stories, Myths, Chants, and Songs of the Kuna Indians
compiled by Joel Sherzer (2004)
With illustrations by the Kuna artist Olokwagdi de Akwanusadup, this engaging book showcases the humor-filled oral traditions of the indigenous Kuna Yala people.

Watch List

Beyond Brotherhood
(2017; drama; directed by Arianne Benedetti)
This moving tale of hardship and redemption revolves around two young siblings forced to live on the streets of Panama City after the death of their parents.

Invasión
(2014; documentary; directed by Abner Benaim)
A deep look at the 1989 US invasion of Panama, with interviews of civilians, soldiers, politicians and even former General Noriega, overthrown in the attack.

Frozen in Russia
(2018; comedy; directed by Arturo Montenegro)
A box-office smash at home, this madcap comedy tells of a fanatic football fan who goes on an absurd quest after Panama qualifies for the World Cup.

The Wind and the Water
(2008; drama; directed by Vero Bollow & Igar Yala Collective)
Indigenous actors star in this beautifully photographed film, which is equal parts love story and cultural collision, as a young Kuna man from the islands seeks his fortune in the big city.

Salsipuedes
(2016; drama; directed by Ricardo Aguilar Navarro & Manuel Rodríguez)
Sent to live in the US at age 10, a young man returns years later to his boyhood *barrio* in Panama, in this feel-good movie about returning home.

The Panama Canal carries about US$270 billion of cargo each year.

A giant of the Latin music scene since the 1970s, from Panama City, Rubén Blades has won nine Grammy awards for his songs featuring Afro-Cuban rhythms and socially progressive themes. He's also an actor, composer, lawyer, politician and activist.

In Panama some of the top reggaeton groups, like La Factoría, are headed by women. Singers Demphra and Joycee led the group to global success before pursuing solo careers, and also helped open doors for other female vocalists.

Playlist

Señorita a Mi Me Gusta Su Style
Los Rabanes
Genre: Ska/Rock

50 ó 100
Entre Nos
Genre: Indie rock

Tu Man
Mecanik Informal
Genre: Salsa

Pedro Navaja
Rubén Blades
Genre: Salsa

Panama 500
Danilo Pérez
Genre: Jazz

Emociones Tantas
Pureza Natural
Genre: Reggae

Todavía
La Factoría
Genre: Reggaeton

Viva Tirado
Los Mozambiques
Genre: Cumbia/Funk

Mi Tierra Te Llora
Margarita Henríquez
Genre: Pop

Panama Verde Panama Red
Cienfue
Genre: Psychedelic/Tropical

Trinidad & Tobago

Read List

A House for Mr Biswas
by V S Naipaul (1961)
The small-town snobbery at the heart of this Trinidadian tale about the often unsympathetic Mohun Biswas can make it a hard read, but a richly rewarding one. It's the novel that brought Naipaul global attention.

The Dragon Can't Dance
by Earl Lovelace (1979)
Carnival, slum life, crime, racial division, post-colonialism and belonging – or not – are all explored to great effect in this thriller set in Port of Spain.

Black Rock
by Amanda Smyth (2009)
Post-colonial, post-WWII Trinidad & Tobago is eloquently evoked in this debut novel about a young, spirited heroine struggling to survive and find herself.

The Golden Child
by Claire Adam (2019)
The desperation of life dictated by poverty in rural Trinidad is beautifully conjured up in this heartbreaking tale about a family forced to make a horrific choice.

Love After Love
by Ingrid Persaud (2020)
Secrets, prejudice and similar dark themes are at the heart of this award-winning novel, but the delightful portraits of Trinidadian characters also make it wickedly funny. The complexity of their shared history gives it a profound depth.

Watch List

Fire Down Below
(1957; drama; directed by Robert Parrish)
The big names – Rita Hayworth, Robert Mitchum and Jack Lemmon, with Albert R Broccoli co-producing – are outshone by the big Caribbean atmosphere and scenery in this love triangle drama.

Bim
(1974; drama; directed by Hugh A Robertson)
Race and class are explored in pitiless depth in this dark drama about a young poor Indian boy, the eponymous Bim.

The Hummingbird Tree
(1992; drama; directed by Noella Smith)
A bittersweet tale about the coming of age and friendship in colonial Trinidad, solidly adapted by the BBC from a book by Trinidadian novelist Ian McDonald.

The Mystic Masseur
(2001; drama; directed by Ismail Merchant)
V S Naipaul's novel about frustrated writer-slash-masseur Ganesh Ramsumair is given the Merchant Ivory treatment in this film exploring issues of Hindu subculture in Trinidad.

Pan! Our Music Odyssey
(2014; documentary; directed by Jerome Guiot & Thierry Teston)
This enthralling documentary with re-enactments offers terrific insight into the human stories behind Panorama, Trinbago's annual battle of the steel pan bands.

Carnival is more than just a festival in Port of Spain, as depicted in Earl Lovelace's *The Dragon Can't Dance.*

Calypso was born in early 20th-century Trinidad as a witty, incisive form of social commentary that not only poked sly or satirical fun at politics, politicians, hypocrites and social climbers, but also told ribald, lewd stories about sex, marriage and infidelity.

Soca rules the road at Carnival time, when the big stars battle it out to see whose anthemic tune will win Carnival's Road March – the song played most often along the parade route. Most years it's Machel Montano, whose insanely catchy rhythms will buzz around in your head for weeks after you've heard them.

Playlist

Jean and Dinah
Mighty Sparrow
Genre: Calypso

Rum and Coca Cola
Lord Invader
Genre: Calypso

Bahia Girl
David Rudder
Genre: Calypso

Burn Dem
Black Stalin
Genre: Calypso

Indrani
Lord Shorty
Genre: Soca

Too Young to Soca
Machel Montano
Genre: Soca

Jahaji Bhai
Brother Marvin
Genre: Chutney soca

Tonite Is De Nite
Brother Resistance
Genre: Rapso

Alegría, Alegría
Daisy Voisin
Genre: Parang

Fire Down Below
Lennon 'Boogsie' Sharpe & Phase II Pan Groove
Genre: Steel pan

USA

Read List

Adventures of Huckleberry Finn
by Mark Twain (1884)
Twain's account of a wandering Huck and his adventures along the Mississippi has been read, loved and critiqued since publication. Often described as 'the great American novel', it's also a wonderfully immediate and exciting read.

The Heart is a Lonely Hunter
by Carson McCullers (1940)
McCullers' deft and affecting debut about the small-town South made her a star at the age of 23, giving voice to characters often neglected in literature.

The Color Purple
by Alice Walker (1982)
Walker's vivid, Pulitzer-winning account follows Celie, an African-American woman from Georgia who suffers at the hands of her father and husband before striking out alone.

Blood Meridian
by Cormac McCarthy (1985)
McCarthy's feverish, brutal anti-Western follows a young man who joins a gang on the lawless Mexican border in the 1840s.

A Visit from the Goon Squad
by Jennifer Egan (2010)
Egan won the Pulitzer for this series of interlinked tales, which feature rock stars, dictators, theft, death and swimming. They range from the 1970s to the present day, but coalesce into a rich, symphonic whole.

Watch List

Singin' in the Rain
(1952; musical; directed by Stanley Donen & Gene Kelly)
This witty, exuberant classic is packed with spectacle and has Kelly and Debbie Reynolds romancing while creating the world's first musical movie.

The Searchers
(1956; Western; directed by John Ford)
Stunning landscapes, a grizzled John Wayne, moral ambiguity and classic set pieces distinguish this search for a kidnapped girl, perhaps the finest Western ever made.

The Godfather
(1972; drama; directed by Frances Ford Coppola)
Lush, sombre and impeccable mafia epic that also offers a fine portrait of 1950s New York, a masterclass in 1970s auteur film-making and a Brando-Pacino double header.

Clueless
(1995; comedy; directed by Amy Heckerling)
The definitive teen movie, full of comedy, angst, satire and excellent clothes. It's based on Jane Austen's Emma, but the vibe is pure 1990s.

Moonlight
(2016; drama; directed by Barry Jenkins)
Jenkins examines sexuality and identity in this sensitive portrayal of an out-of-place African American teen growing up in Miami. It's a phenomenal piece of film-making that hits you in the tear ducts and the gut.

Steamboats like the *Natchez* plied the Mississippi in the time of Mark Twain.

Some of the US's finest songs examine America itself, often in a mix of patriotism and anger. *Born in the USA* is a classic example, its bleak lyrics driven forward by a fist-pumping synth riff.

Some of the US's greatest music has been associated with specific cities. Howlin' Wolf and many blues acts came from Chicago, Detroit has Motown and techno, LA hair-metal, San Francisco psychedelic rock and Atlanta Southern hip-hop and trap.

Playlist

The Star Spangled Banner
Jimi Hendrix
Genre: Rock

This Land is Your Land
Woody Guthrie
Genre: Folk

Ancestral Home
R Carlos Nakai
Genre: Instrumental

Born in the USA
Bruce Springsteen
Genre: Rock

Strange Fruit
Billie Holiday
Genre: Jazz

Smokestack Lightnin'
Howlin' Wolf
Genre: Blues

Yankee Doodle
38th Army Band
Genre: March

Empire State of Mind
Jay-Z ft Alicia Keys
Genre: Hip-hop

Like a Prayer
Madonna
Genre: Pop

In the Ghetto
Elvis Presley
Genre: Pop

IN DEPTH

Jazz

from THE UNITED STATES

Jazz is the music that gave birth to an age. It encompasses everything from joyful foxtrot and twitching swing to soaring improvisation and jazz rap. And it began 150 years ago, in New Orleans.

Here, formerly enslaved people adapted reed, horn and string instruments to play their own African-influenced music, and this cross-pollination produced a stream of innovative sounds. In the 1890s came ragtime, so-called because of its 'ragged,' syncopated rhythms. Dixieland jazz, with its addictive countermelodies, was next. With his distinctive vocals and talented improvisations, trumpeter Louis Armstrong led to the solo becoming an integral part of jazz in the 1920s.

© YALE JOEL / THE LIFE PICTURE COLLECTION VIA GETTY IMAGES

© KRIS DAVIDSON / LONELY PLANET

The 1920s and '30s are known as the Jazz Age, in which music was a key part of the flowering of African American culture during New York's Harlem Renaissance. Swing – an urbane, big-band style – swept the country, led by Duke Ellington and Count Basie. Singers Ella Fitzgerald and Billie Holiday combined jazz with its southern sibling, the blues.

After WWII, a new crop of bebop musicians reacted against the smooth melodies of big-band swing, including Charlie Parker and Dizzy Gillespie, before hard-bop, free jazz and fusion further ripped up the rulebook under the influence of players such as Miles Davis and Charles Mingus. Today, jazz may no longer be commercially dominant, but it's still producing artists such as Kamasi Washington, who mixes Latin, psychedelic and funk influences into an urgent sound that speaks to jazz's glorious past and to the struggles of the present.

Left to right: Ella Fitzgerald performing at Mister Kelly's in Chicago; Preservation Hall in New Orleans, a crucible of jazz; Miles Davis, jazz iconoclast, at Newport Jazz Festival.

OCEA

ANIA

Australia

Read List

Rabbit-Proof Fence
by Doris Pilkington Garimara (2007)
The true story of three mixed-race Aboriginal sisters who were part of the so-called Stolen Generation, taken from their parents in 1931 as attempts were made to integrate Aboriginal children into white families. It offers insights into the plight of Australia's original landowners.

The Songlines
by Bruce Chatwin (1986)
Chatwin traces the songlines, the invisible, indigenous pathways across Australia that are mapped out by traditional Aboriginal art, touching on existential themes.

True History of the Kelly Gang
by Peter Carey (2000)
This extraordinary feat of imagination tells the story of Aussie anti-hero Ned Kelly and the landscape of northeast Victoria in the bushranger's own vernacular. You can visit Kelly's lookout and hideouts today.

The Secret River
by Kate Grenville (2005)
As much a story of the Australian landscape, 'a place out of a dream', as a tale of 19th-century convicts and the colonisation of Australia.

Sand Talk: How Indigenous Thinking Can Save the World
by Tyson Yunkaporta (2020)
Yunkaporta's scientific treatise offers an alternative template for how to live sustainably, equal parts practical and philosophical.

Watch List

Two Hands
(1999; comedy; directed by Gregor Jordan)
A young Heath Ledger stars as Jimmy, who finds himself in debt to a Sydney gangster. Meanwhile, two street kids start a shopping spree when they find some missing money.

Walkabout
(1971; drama; directed by Nicholas Roeg)
Two city-bred children are abandoned to contend with Australia's harsh wilderness. They are saved by an Aboriginal boy who shows them how to survive and, in the process, underscores the disharmony between nature and modern life.

The Hunter
(2011; drama; directed by Daniel Nettheim)
The tension between hardscrabble Australian livelihoods and care for a fragile environment comes to the surface in a Tasmania-set drama with Willem Dafoe – and a guest appearance from an extinct thylacine.

The Castle
(1997; comedy; directed by Rob Sitch)
'Tell him he's dreaming': this is a story of the suburban dream that will make sense of important Australian catchphrases. Be sure to visit serene Bonny Doon.

The Adventures of Priscilla, Queen of the Desert
(1994; musical; directed by Stephan Elliott)
Follow drag queens Hugo Weaving and Guy Pearce in a bus across the Outback in this LGBTQ classic.

The Tasmanian bush, once home to the thylacine, is the setting for *The Hunter*, starring Willem Dafoe.

Playlist

Black And Blue
Chain
Genre: Blues

Jailbreak
AC/DC
Genre: Rock

I Am Woman
Helen Reddy
Genre: Folk

Bradman
Paul Kelly
Genre: Rock

Paul Kelly is perhaps the bard of Australia, who has been singing heartfelt tales since the 1970s. *Bradman* is his paean to cricketer Don Bradman, Australia's greatest batsman.

Black Fella White Fella
The Warumpi Band
Genre: Rock

The Best of Both Worlds
Midnight Oil
Genre: Pop/Rock

Djarimirri
Geoffrey Gurrumul Yunupingu
Genre: Indigenous

Depreston
Courtney Barnett
Genre: Indie

Cattle and Cane
The Go-Betweens
Genre: Pop

Let It Happen
Tame Impala
Genre: Pyschedelic pop

Tame Impala is just one of the projects of polymathic, Perth-based musician Kevin Parker. He's part of an Australian neo-psychedelic scene that also includes Pond and King Gizzard and the Lizard Wizard.

New Zealand

Read List

The Bone People
by Keri Hulme (1984)
New Zealand's first Booker Prize winner is a nuanced and at times harrowing novel involving love, abuse, violence and Māori spirituality, set on the West Coast of the South Island.

Bulibasha: King of the Gypsies
by Witi Ihimaera (1994)
The country's most accomplished writer is at the top of his game in this epic novel of family relationships in a rural Māori community. There's a great plot twist, too.

Dogside Story
by Patricia Grace (2001)
Another tale of rural Māori life and family conflict set around the dawn of the new millennium, in the first place in the world to see the sun rise.

Rangatira
by Paula Morris (2011)
Morris retraces the steps of one of her ancestors in this novel, based on the true story of a *rangatira* (Māori chief) who travelled to London to meet Queen Victoria.

The Luminaries
by Eleanor Catton (2013)
Set on the South Island's West Coast during the 19th-century gold rush, this Man Booker Prize–winner has an intriguing structure and a rollicking, if complicated, plot.

Watch List

The Piano
(1993; drama; directed by Jane Campion)
Local actors Sam Neill and Anna Paquin (in her Oscar-winning role) star alongside Holly Hunter and Harvey Keitel in this grim story of a loveless marriage, set in the early colonial era.

Heavenly Creatures
(1994; drama; directed by Peter Jackson)
Before Sir Peter's knighthood and Oscar he made quirky arthouse flicks like this, combining whimsical magic realism with the real events of a famous 1950s Christchurch murder case.

Once Were Warriors
(1994; drama; directed by Lee Tamahori)
This powerful tale of domestic violence, rape and incest in a poverty-stricken South Auckland family is a harrowing adaptation of an Alan Duff novel.

Whale Rider
(2002; drama; directed by Niki Caro)
This wonderful adaptation of Witi Ihimaera's story of female empowerment within the bounds of traditional Māori society is sweet, funny and uplifting. It won its young star, Keisha Castle-Hughes, an Oscar nomination.

Hunt for the Wilderpeople
(2016; comedy drama; directed by Taika Waititi)
Like all of Waititi's movies, *Hunt for the Wilderpeople* is quirky, hilarious and immersed in Kiwi culture. It's set in the New Zealand wilderness and wins for its huge heart.

A Māori man aboard his *waka*, a traditional war canoe; indigenous culture continues to inspire contemporary stories.

Written by the warrior chief Te Rauparaha around 1820, *Ka Mate* is a dramatic rumination on life and death. It's the most famous of all *haka* (ceremonial dances of challenge) due in part to its adoption by the All Blacks, NZ's national rugby union team, as a pre-match ritual since 1905.

The Chills emerged from Dunedin in the 1980s and exemplified what came to be known as 'the Dunedin sound'. The scene was hugely influential on '90s alternative music, with bands such as R.E.M. citing it as an inspiration.

Playlist

Oceania

Ka Mate
The All Blacks
Genre: Kapa haka

Pokarekare Ana
Kiri Te Kanawa
Genre: Kapa haka/Opera

Six Months in a Leaky Boat
Split Enz
Genre: Rock

Poi E
Patea Maori Club
Genre: Pop/Kapa haka

Land of Plenty
OMC
Genre: Hip-hop

Not Many
Scribe
Genre: Hip-hop

Heavenly Pop Hit
The Chills
Genre: Indie

Team
Lorde
Genre: Pop

Haere Mai Rā/Sway
Bic Runga
Genre: Pop

Kua Kore He Kupu/ Soaked
Benee
Genre: Pop

IN DEPTH

Māori Renaissance

from NEW ZEALAND

When filmmaker Taika Waititi dedicated his 2020 Oscars win to artsy indigenous kids worldwide it was just the latest example of the confident role that Māori artists are playing on the world stage. In Aotearoa/NZ itself, Māori culture has a huge influence on every corner of the creative arts – yet this wasn't always the case.

Traditional Māori culture always valued the arts, as expressed in intricate carving, weaving and kapa haka (song and dance). Yet colonialism saw Māori creativity relegated to a supporting role, trotted out for tourists and rugby matches. While some Māori artists excelled in European art forms (opera singer Dame

© LEONARDO CENDAMO / GETTY IMAGES

© DAVID CROTTY / GETTY IMAGES

Kiri Te Kanawa, for example), it wasn't until the latter part of the 20th century that a new wave of creatives emerged, telling Māori stories in Māori ways. This is especially apparent in the field of literature, with many of the country's most acclaimed writers boasting Māori heritage, notably Witi Ihimaera, Patricia Grace, Keri Hulme, Alan Duff, Paula Morris and Becky Manawatu. The same is true in the fields of filmmaking, acting, music and the visual arts.

A welcome recent trend is the increased use of the Māori language in popular song, with artists such as Bic Runga and Benee releasing versions of their hits in Māori.

Left to right: New Zealand novelist and children's author Patricia Grace; Oscar-winning director (and writer and actor) Taika Waititi; musician Bic Runga.

Papua New Guinea

Read List

My Childhood in New Guinea
by Paulias Matane (1972)
One of PNG's most prolific authors, Matane writes movingly of his childhood on New Britain island, providing insight into the traditions and rituals of village life.

The Crocodile
by Vincent Eri (1970)
Evoking a spiritual world full of myths and sorcery, *The Crocodile* follows a young Papuan villager living with colonial encroachment and the upheavals of WWII.

A Thousand Coloured Dreams
by Josephine Abaijah (1991)
Although a work of fiction, this story is based largely on the author's life, describing encounters with colonialism and oppression faced by so many women. In 1972 Abaijah became the first woman elected to PNG's parliament.

Ten Thousand Years in a Lifetime
by Albert Maori Kiki (1968)
An astonishing account of growing up amid the semi-nomadic highland Kukukuku people, with descriptions of Kiki's initiation ceremony, encounters with Western outsiders, and his later political awakening.

Throwim Way Leg
by Tim Flannery (1998)
Pidgin for going on a journey, *Throwim Way Leg* documents the astonishing adventures and scientific discoveries of Australian zoologist Tim Flannery, over 15 research excursions to the island beginning in 1981.

Watch List

Lukim Yu
(2016; drama; directed by Christopher Anderson)
This film shows the struggles and ambitions of a group of young friends living in Port Moresby, while addressing gender equality, violence, sorcery and arranged marriages.

The Coconut Revolution
(2001; documentary; directed by Dom Rotheroe)
One of the world's first eco-revolutions is the subject of this award-winning documentary, following the indigenous uprising on Bougainville Island incited by devastating foreign-owned mining operations.

Aliko & Ambai
(2017; drama; directed by Diane Anton & Mark Eby)
The story of two young women growing up in the eastern highlands, seeking a brighter future amid forced marriages, domestic violence and tribal conflicts.

I'm Moshanty. Do You Love Me?
(2019; documentary; directed by Tim Wolff)
This compelling portrait of a talented singer and transgender human-rights activist shines a light on the bigotry and violence facing the LGBTIQ+ community.

Mr Pip
(2012; drama; directed by Andrew Adamson)
Shot largely on location, *Mr Pip* tells of the harrowing struggles of a group of young students and their inspiring teacher during the time of Bougainville's civil war.

Papua New Guinea's unique wildlife, including birds of paradise, is described in *Throwim Way Leg*.

Singing in Kuanua and Tok Pisin, George Telek is one of PNG's most successful artists and has garnered international awards for his ground-breaking albums, featuring three-part harmonies backed by acoustic guitars and traditional drums.

Straddling two cultures, Ngaiire Joseph was born and raised in Papua New Guinea but moved to Australia at the age of 16. She fuses traditional Melanesian rhythms with lush vocals on her polished 2013 album *Lamentations*.

Playlist

Oceania

Namilai
Sanguma
Genre: Traditional

Midal
Telek
Genre: Traditional/Pop

Motomse
Sambra Aikit
Genre: Funk

PNG Pom City
Naka Blood
Genre: Hip-hop

Mangi Tolai
Nathan Nakikus ft Anslom Nakikus
Genre: Reggae

Around
Ngaiire
Genre: Neo soul

Sweet Mekeo
Sprigga Mek
Genre: Hip-hop

Island Jewel
O-Shen
Genre: Reggae

Inagwe
Richard Mogu
Genre: Folk

Wanchef
Airileke
Genre: Fusion

Samoa

Read List

Island Nights' Entertainments
by Robert Louis Stevenson (1893)
Scottish-born Stevenson spent the last four years of his life in Samoa, where he was affectionately known as Tusitala (storyteller). This collection of island-based stories was his last major work.

Where We Once Belonged
by Sia Figiel (1996)
Village life is captured through the eyes of an adolescent girl in this novel. It's full of humour and sharp observation, yet also covers violent and disturbing themes.

The Mango's Kiss
by Albert Wendt (2003)
Samoa's pre-eminent writer deals with the *fa'a Samoa* (Samoan worldview) bumping against *palagi* (European) ideas in this engrossing novel, which follows the fortunes of a family in the early 20th century.

Fast Talking PI
by Selina Tusitala Marsh (2009)
The PI of the title stands for Pacific Islander, and this collection of poems from the former New Zealand Poet Laureate riffs on themes of Pasifika identity.

Telesā – The Covenant Keeper
by Lani Wendt Young (2011)
Targetted at young adult readers and set mainly in Samoa, the *Telesā* novels combine Samoan mythology with teen romance. This is the first book in the ground-breaking series.

Watch List

Flying Fox in a Freedom Tree
(1989; drama; directed by Martyn Sanderson)
Adapted from a novel by Albert Wendt, this film follows the story of a young man, Pepe, as he rebels against his father's rigid beliefs and greed.

Sione's Wedding
(2006; comedy; directed by Chris Graham)
Although set in Auckland, this hugely successful comedy couldn't be more Samoan. It follows the exploits of four lads banned from attending the titular wedding – unless they find girlfriends first.

The Orator
(2011; drama; directed by Tusi Tamasese)
Steeped in Samoan culture, this critically acclaimed movie was the first feature film set in Samoa, with a Samoan cast, telling a Samoan story in the Samoan language.

Three Wise Cousins
(2016; comedy; directed by Stallone Vaiaoga-Ioasa)
Adam, a New Zealand-born Samoan, travels to his ancestral homeland to learn how to be a 'real Pacific Island man' so he can impress the object of his affections.

Moana
(2016; animated; directed by Ron Clements & John Musker)
This heart-warming pan-Polynesian tale features the voice of the world's most famous Samoan, Dwayne 'the Rock' Johnson, as demigod Maui. The music was co-written by Samoan-born Opetaia Foa'i.

Dwayne Johnson, Samoan superstar and the voice of an animated demigod in *Moana*.

This dancefloor banger was a massive hit for New Zealand-born Samoan rapper Savage, reaching multiplatinum status in New Zealand, Australia and the US after it featured in the movie *Knocked Up*.

Much of the music for Disney's *Moana* was written by Opetaia Foa'i, including this uplifting anthem. It has become a live staple for Foa'i's long-standing pan-Polynesian band Te Vaka; check out their version with Orchestra Wellington on YouTube.

Playlist

Oceania

We are Samoa
Jerome Grey
Genre: Easy listening

Samoa Ea
The Yandall Sisters
Genre: Easy listening

Swing
Savage
Genre: Hip-hop

Samoa Matalasi
The Five Stars
Genre: Easy listening

Screems From Da Old Plantation
King Kapisi
Genre: Hip-hop

I Am Moana (Song of the Ancestors)
Te Vaka
Genre: Musical

Su'amalie / Aint Mad At You
Tha Feelstyle
Genre: Hip-hop

FRESH
Scribe
Genre: Hip-hop

Nesian 101
Nesian Mystik
Genre: Hip-hop

Malu Afiafi
Ben Vai
Genre: Easy listening

Timor-Leste

Read List

The Crossing
by Luís Cardoso (2000)
A lyrical, semi-autobiographical novel based on Cardoso's experiences growing up in Timor-Leste before the Indonesian invasion, and his life as an exile in Portugal.

Funu: The Unfinished Saga of East Timor
by José Ramos-Horta (1987)
The co-recipient of the 1996 Nobel Peace Prize (and former Prime Minister and former President of Timor-Leste) sheds light on the world's indifference to the horrors of Indonesia's aggression.

Shooting Balibo
by Tony Maniaty (2009)
In this moving memoir, Australian journalist Maniaty returns to Timor-Leste with the crew of the film *Balibo*, 30 years after he was there with the Balibo Five, to lay his ghosts to rest.

A Dirty Little War
by John Martinikus (2001)
An eyewitness story of Indonesia's sustained campaign of terror from 1997 to 1999 against one of Australia's closest neighbours, written by an acclaimed Australian journalist.

Beloved Land: Stories, Struggles, and Secrets from Timor-Leste
by Gordon Peake (2013)
Based on four years spent living in the country, UK writer Peake explores the daunting hurdles the people of Timor-Leste must overcome to build a nation from scratch.

Watch List

Balibo
(2019; drama-thriller; directed by Robert Connolly)
Based on the nonfiction book by Jill Jolliffe, this hit film recounts the investigation of the 1975 murders of the journalists in Timor-Leste known as the Balibo Five.

Beatriz's War
(2013; drama; directed by Luigi Acquisto & Bety Reis)
The first full-length feature film to be produced by Timor-Leste revolves around the life of Beatriz, a young East Timorese woman separated from her new husband as the Indonesian invasion of Timor-Leste begins.

Abdul & José (The Stolen Child)
(2017; documentary; directed by Luigi Acquisto & Lurdes Pires)
The moving true story of 'José' Abdul Rahman, who was taken by Indonesians in 1979 as a boy, and returned to Timor-Leste 36 years later to reconnect with his family.

The Stolen Children of Timor-Leste
(2020; documentary; produced by Foreign Correspondent/ABC)
The ABC's Indonesia Correspondent Anne Barker follows a group of Timor-Leste-born adults as they return to reunite with their families.

East Timor – Birth of a Nation
(2002; documentary; directed by Luigi Acquisto & Andrew Sully)
Each programme in this two-part series tells the powerful story of a remarkably resilient East Timorese individual.

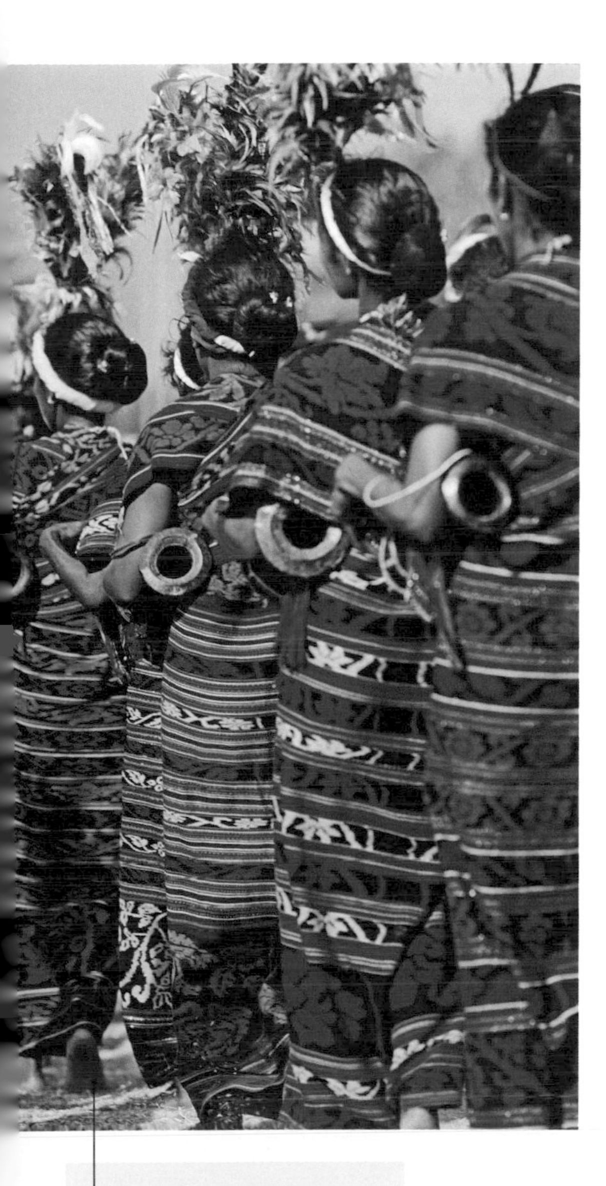

Traditionally attired women celebrating independence from Indonesia in Timor-Leste.

East Timorese poet Francisco Borja da Costa, who penned national anthem *Pátria* and the pro-independence Fretilin military anthem *Foho Ramelau*, was executed by Indonesian forces on the second day of the occupation in 1975.

Hakotu Ba (*Please Decide*) was commissioned by the UN to encourage East Timorese to register for their right to vote in the 1999 referendum for independence.

Playlist

Oceania

Oh Hele Le
Ego Lamos
Genre: Traditional folk

Liberdade
Dili Allstars
Genre: Folk

Foho Ramelau
Francisco Borja da Costa
Genre: Traditional folk

Pátria
Francisco Borja da Costa
Genre: Patriotic

Hakotu Ba
Lahane
Genre: Traditional folk

Ego Lamos
Balibo
Genre: Folk

Timor Hau Hadomi O
5 Do Oriente
Genre: Pop

Labarik Lakon (Stolen Children)
Joviana Guterres
Genre: Folk

Liberdade
Galaxy
Genre: Reggae/Hip-hop

Kolele Mai
Midnight Oil
Genre: Traditional folk

AS

IA

Azerbaijan

Read List

Ali and Nino
by Kurban Said (1937)
While the 2016 film adaption was well received, the romance between Azerbaijani Muslim Ali and Georgian Christian Nino in Baku in the wake of WWI is arguably best captured in the classic novel that inspired it.

The Orphan Sky
by Ella Leya (2015)
Set in Communist Baku in the late 1970s, the Baku-born US author's debut novel reveals a young pianist's struggle to decide whether to betray her country or her heart.

Classic Poetry of Azerbaijan: An Anthology
by Paul Smith (2013)
One of the few anthologies in English of the greatest poets of Azerbaijan in the classic period, between the 11th and 17th centuries. Contributions range from the great Nizami Ganjavi to lauded female poet Mahsati.

Solar Plexus, A Baku Saga in Four Parts
by Rustam Ibragimbekov (1996, trans 2012)
A compelling saga of family and friendship, love and betrayal, set against the backdrop of a rapidly changing Baku as Azerbaijan grapples with its emergence into a post-Soviet world.

Days in the Caucasus
by Banine (1945; trans 2019)
Also called *Caucasian Days*, this memoir recalls its author's wealthy upbringing in oil-boom Baku in the early 20th century before she escaped to Persia, Georgia and eventually Paris.

Watch List

Pomegranate Orchard
(2017; drama; directed by Ilgar Najaf)
After a sudden departure 12 years earlier, a man returns to his humble family pomegranate farm to find that the deep emotional scars he left behind cannot be easily erased.

Nabat
(2014; drama; directed by Elchin Musaoglu)
The heartbreaking story of a woman living in the disputed territory of Nagorno-Karabakh, who befriends a wolf following the death of her son and husband.

On Distant Shores
(1958; war drama; directed by Tofig Taghizade)
Portrays the life of legendary Azerbaijani guerrilla Mehdi Huseynzade, who fought the Nazis in present-day Italy and Slovenia, with a sprinkling of Soviet-era propaganda.

Holy Cow
(2015; documentary; directed by Imam Hasanov)
A man's dream of bringing a European cow into his Azeri village unsettles the conservative community that wants to keep its secular traditions intact.

Buta
(2011; drama; directed by Ilgar Najaf)
In a poor mountain village, young boy Buta befriends an elderly man. Meanwhile, in a time-honoured Azeri tradition, Buta's grandmother weaves a special carpet in memory of Buta's mother.

Ancient and futuristic meet in Baku, the Azeri capital and setting for *Ali and Nino* and *Days in the Caucasus*.

Kazakhstan-born, Russian rapping Khalib, who has Azeri heritage, has become one of the most popular modern artists of the region. His 2015 hit *You Are Like a Whole Universe* alone has clocked up 17.5 million streams (and counting) on Spotify.

Pop duo Ell & Niki secured their spot in the Eurovision hall of fame by being the first Azeri group to win the competition with their hit song *Running Scared* in 2011.

Playlist

Baku
Alim Qasimov
Genre: Folk

Bayatı-Şiraz
Kenan Bayramlı
Genre: Traditional folk

You Are Like a Whole Universe
Jah Khalib
Genre: Hip-hop/Rap

Sari Gelin
Various artists
Genre: Traditional folk

Men Gülem
Brillant Dadaşona ft Alihan Samedov
Genre: Folk/Pop

Ay Giz
Rahman Asadollahi
Genre: Traditional folk

Running Scared
Ell & Nikki
Genre: Pop

Coffee from Colombia
Aygün Kazımova ft Snoop Dogg
Genre: Pop

Drip Drop
Safura Alizadeh
Genre: Pop

I'm Free
Eldar Gasimov
Genre: Dance

Bangladesh

Read List

Ekattorer Dinguli (Days of 71)
Jahanara Imam (1986)
Diary turned-bestseller, this harrowing first-hand account of Bangladesh's brutal Liberation War by a local mother, teacher and socialist activist covers the tumultuous events of 1971 that led to the nation's independence.

Lajja (Shame)
Taslima Nasrin (1993, translated 1997)
This controversial novel is based around the 1992 Ayodhya religious riots in India, portrayed through the experiences of a Hindu family living in Bangladesh. The book is still banned today, and the author remains in exile.

Sultana's Dream
Rokeya Sakhawat Hossain (1905)
A ground-breaking feminist sci-fi novel written in 1905, where the traditional Islamic practice of purdah is reversed – women run everything and men live in secret.

A Golden Age
Tahmima Anam (2007)
The acclaimed debut work of Bangladeshi-born Tahmima Anam is this page-turner set during the Liberation War, covering nationalism and family loyalties.

Banker to the Poor
Dr Muhammad Yunus (2003)
Written by economist, social entrepreneur and 2006 Nobel Peace Prize joint winner, Muhammad Yunus' autobiography begins with his first-hand experiences of witnessing mass poverty in Chittagong, an event that inspired him to found the Grameen Bank for the poor.

Watch List

The Clay Bird
(2002; drama; directed by Tareque Masud)
The only Bangladeshi film to feature at Cannes is this award-winning independent production, based on a true story exploring the director's childhood. It's set in a small village during the years leading up to the Liberation War.

Made in Bangladesh
(2019; drama; directed by Rubaiyat Hossain)
This honest and beautifully shot contemporary drama premiered at Toronto's International Film Festival. It centres around a Dhaka sweat shop and covers issues relating to women's rights.

Aguner Poroshmoni
(1994; drama; directed by Humayun Ahmed)
Cleaning up in Bangladesh's National Film Awards, this cult classic is set during the Liberation War. It's based on a book written by director Ahmed, who is also a novelist.

Lalsalu
(2001; tragicomedy; directed by Tanvir Mokammel)
Based on a classic novel of the same name, this story focuses on a disreputable Islamic cleric who appears suddenly in a remote village to convince locals the town is the burial place of a famed holy man.

Ironeaters
(2007; documentary; directed by Shaheen Dill-Riaz)
This German-Bangla documentary explores Chittagong's controversial ship-breaking industry. It sheds light on the dangers, deplorable working conditions and the exploitation of labourers.

Dhaka, a city of nine million people, is the setting for the movie *Made in Bangladesh*.

No list on Bangladesh is complete without Bengali polymath Rabindranath Tagore. While he mostly lived in India, he was born in Jessore and composed Bangladesh's national anthem, his ode to 'Mother Bengal'.

Known as both a pioneer of Bangladeshi rock and a freedom fighter during the Liberation War, Azam Khan's music has a counter-culture beat. His song *Ore Saleka, Ore Maleka* became a youth anthem for the birth of a new nation.

Playlist

Amar Sonar Bangla
Rabindranath Tagore
Genre: Bengali folk

Ami Jaar Nupurer
Firoza Begum
Genre: Traditional

Krishno
Habib Wahid ft Kaya
Genre: Electronic

Bondho Janala
Shironamhin
Genre: Rock/Pop

Ami Kul Hara Kolongkini
Shah Abdul Karim
Genre: Folk

Jaat Gelo Jaat Gelo Bole
Farida Parveen
Genre: Traditional

Ore Saleka, Ore Maleka
Azam Khan (Uchcharon)
Genre: Rock/Pop

Jibon Dhara
Warfaze
Genre: Rock

Amar Gaye Joto Dukkho Shoy
Bari Siddiqui
Traditional

Bangadesh
George Harrison
Genre: Rock/Pop

Cambodia

Read List

First They Killed My Father
by Loung Ung (2000)
A powerful memoir from the daughter of a Cambodian government official, whose comfortable life was ripped apart as the Khmer Rouge forcibly remodelled Cambodia into an agrarian economy.

Reamker
unknown authors (7th to 18th centuries)
The Cambodian retelling of the Indian epic, the *Ramayana*, reimagines Hindu legends with Buddhist sensibilities, reflecting the conversion of Khmer society to Buddhism in the 13th century.

Stay Alive, My Son
by Pin Yathay (1987)
The sole survivor from an educated family forced into the fields under Khmer Rouge rule, Pin Yathay's powerful autobiography is about fear, loss and resilience.

Cambodian Folk Stories from the Gatiloke
by Muriel Paskin Carrison & Kong Chhean the Venerable (1993)
A collection of Khmer folk tales originally compiled by scholar Ukna Suttantaprija Ind in 1921, revealing the world as taught by Buddhist monks in the 20th century.

Sophat
by Rim Kin (1942)
This sentimental, layered tale of love, betrayal and loss was hailed as the first Khmer romance novel; it's yet to be translated into English, but French translations are not hard to find.

Watch List

Don't Think I've Forgotten: Cambodia's Lost Rock & Roll
(2014; documentary; John Pirozzi)
An uplifting, sometimes haunting, joint project by US director John Pirozzi and Cambodian academic LinDa Saphan, exploring the rise, destruction and resurgence of Cambodian popular music.

Tum Teav
(2003; romantic drama; directed by Fan Sam Ang)
A dramatisation of one of Cambodia's most popular folk stories, exploring themes of forbidden love and overbearing parents. There's an air of *Romeo & Juliet* tragedy about the ending.

Rice People
(1994; drama; directed by Rithy Panh)
As Cambodia rebuilds, a family tries to relearn how to grow rice, after years relying on sacks of rice from the United Nations; a film full of honest realism.

Khmoach Daoem Chek Chvia
(2004; horror; directed by Heng Tola)
In '*Ghost Banana Tree*', a man is haunted by a vengeful ghost, who enters his home by climbing the banana tree planted close to the walls – a taboo in Khmer culture.

The Missing Picture
(2013; documentary; directed by Rithy Panh)
A remarkable exploration of the Khmer Rouge years, using archive footage and hand-moulded clay figurines, which adds to, rather than diminishing, the horror.

The Tuol Sleng Genocide Museum is s reminder of the Khmer Rouge's atrocities.

Legendary 1960s and '70s singer Ros Sereysothea disappeared during the Khmer Rouge era, and most master recordings of her work were destroyed. Luckily, some vinyl records survived and were reissued when her sister, Ros Saboeut, gathered musicians to rebuild Cambodia's music industry.

The Cambodian Original Music Movement grew out of sessions at Show Box, a community arts hub in Phnom Penh. At the heart of the movement is the search for a uniquely Cambodian sound, a reaction to the tradition of borrowing melodies from Western music.

Playlist

Champa Battambang
Sinn Sisamouth
Genre: Popular jazz

Pka Proheam Rik Popreay
Laura Mam
Genre: Pop

Cry Loving Me
Ros Sereysothea
Genre: 1960s rock

Kam Knea Doch Chkae
Doch Chkae
Genre: Metal

Dey Khmer
Kea Sokun
Genre: Rap

Je Te Quitterai
Mol Kamach & Baksey Cham Krong
Genre: 1960s pop

Diamond
Adda Angel
Genre: Cambodian Original Music Movement

Home for Dinner
Nikki Nikki
Genre: Pop

Violent Reform
Nightmare AD
Genre: Punk

Have Visa, No Have Rice
Kak Channthy & the Cambodian Space Project
Genre: Alt rock

China

Read List

Dream of the Red Chamber
by Cao Xueqin (1791, trans 1973-80)
Offering detailed insights into upper-class society, this 18th-century masterpiece tells the story of the rise and fall of an aristocratic family during the Qing dynasty.

Please Don't Call Me Human
by Wang Shuo (2000)
Written by one of China's bestselling authors, this satire novel explores contemporary urban Chinese life. Fast paced, humorous and a satisfying, climatic ending.

Diary of a Madman
by Lu Xun (1918)
Written by the father of modern Chinese fiction – and considered trailblazing for its use of colloquial language – is this haunting short story of a man who fears cannibals are on the prowl within his village.

Soul Mountain
by Gao Xingjian (1989, trans 2000)
This semi autobiographical book written by dissident writer Gao Xingjian won the Nobel Prize for Literature in 2000. It's the story of his journey along the Yangzi following his misdiagnosis with cancer.

Shanghai Baby
by Wei Hu (1999, trans 2001)
Written at the age of 27, Wei Hu's controversial bestselling debut novel draws on her own experiences from Shanghai's nightlife scene. It was banned and publicly burnt because of its references to sex and drugs.

Watch List

Yellow Earth
(1984; drama; directed by Chen Kaige)
Signifying the beginning of a golden era in China filmmaking, this internationally acclaimed movie set in the late 1930s follows the journey of a CCP soldier to remote Shaanxi in search of folk songs for the army.

Ju Dou
(1990; drama; directed by Zhang Yimou)
The first Chinese movie nominated for the Academy Awards is this beautifully shot film that tells the harrowing story of a dysfunctional relationship set in rural China.

Farewell My Concubine
(1993; drama; directed by Chen Kaige)
This enthralling historical drama featured in the New York Times 'Best 1000 Films of all Time'. It's about the complex relationship between two Peking opera actors in a period spanning the 20th century.

The Blue Kite
(1993; drama; directed by Tian Zhuangzhuang)
Another film banned by the Chinese censors, this heartbreaking story of a Beijing family during the Cultural Revolution. A bleak appraisal of political repression at the height of Mao's Great Leap Forward.

Still Life
(2006; drama; directed by Jia Zhangke)
This Venice Film Festival–winning movie is set to the backdrop of small-town China in an area marked for demolishment to make way for the construction of the Three Gorges Dam.

Asia

Shanghai's skyline; Wei Hu writes about the city in her novel *Shanghai Baby*.

Playlist

Man Jiang Hong (A River of Blossoms)
SMZB
Genre: Punk

Love Letter to myself
Faye Wong
Genre: C-Pop

Nothing to My Name
Cui Jian
Genre: Rock

Performed by China's 'Father of Rock', this powerful and rousing ballad by Cui Jian was adopted by students as a political anthem during the Tiananmen Square protests in 1989. It remains one of the most influential songs in Chinese rock history.

Zhong Nan Hai
Carsick Cars
Genre: Indie rock

Little Apple
Chopstick Brothers
Genre: C-Pop

Welcome to Beijing
Yin Ts'ang
Genre: Rap

Two Springs Reflecting the Moon (Er Quan Ying Yue)
Abing
Genre: Traditional/Folk

This melodic piece played using an *erhu* (a traditional two-stringed bowed instrument) is the signature song from Abing – a blind musician regarded as one of China's most famous of the 20th century.

The Moon Represents My Heart
Teresa Teng
Genre: Pop

One Man War
PK 14
Genre: Post-punk

Eagle
Mamer
Genre: Kazakh

India

Read List

The Autobiography of an Unknown Indian
by Nirad C Chaudhuri (1951)
Chaudhuri was born in 1897, and his thoughtful account tells of his youth in Bengal and Calcutta, as well as exploring Indian culture and British influence as the country moves to independence.

Midnight's Children
by Salman Rushdie (1981)
Rushdie's Booker-winning masterpiece is at once fantastical and hard-hitting, epicly ambitious and genuinely fun. Narrator Saleem wrestles with the impacts of partition, nasal congestion and magical gifts.

A Fine Balance
by Rohinton Mistry (1995)
The state of emergency declared by Indira Gandhi in 1975 is at the heart of Mistry's beautiful and sad account of post-independence India, which follows four very different protagonists.

The God of Small Things
by Arundhati Roy (1997)
Roy's debut uses lush prose and time shifts to tell of the separate lives of two Keralan twins, separated in their childhood, and their extended family.

White Tiger
by Aravind Adiga (2008)
A boy moves from rural poverty to urban wealth in Adiga's thrilling and merciless portrait of a modern India tainted by corruption, murder and rampant inequality.

Watch List

Mother India
(1957; drama; directed by Mehboob Khan)
Khan's hugely influential epic follows a mother's struggle to raise her family alone. This melodramatic, politically charged piece of social realism played non-stop in Indian cinemas for three decades.

Sholay
(1975; thriller; directed by Ramesh Sippy)
It sank on first release, but this multigenre ('masala') Bollywood western has become a huge favourite. Two criminals are hired to capture a bandit in a film that brims with battles, dance and broad humour.

Dilwale Dulhania Le Jayenge
(1995; romantic comedy; directed by Aditya Chopra)
'*The Big-Hearted Will Take the Bride*' is a much-imitated rom-com, with Shah Rukh Khan and Kajol shining as two Indians who fall in love while holidaying in Europe.

3 Idiots
(2009; drama; directed by Rajkumar Hirani)
Two men unite to seek their inspirational college friend Rancho, who disappeared a decade ago. The most successful Bollywood film ever, it mixes laughs and plot twists with compassion.

Dangal
(2016; drama; directed by Nitesh Tiwari)
A former wrestler trains his two daughters in a slick sporting epic that approaches family tussles and literal grappling with an equal and inspiring zest.

Rajkumar Hirani, director of the Bollywood movie *3 Idiots*, was born in Nagpur.

India's classical traditions are Hindustani (from North India) and Carnatic (from South India). At the heart of both are *raga* (melodic structure) and *tala* (rhythm), with numerous combinations of the two possible – Hindustani music in particular encourages improvisation.

Goa trance was born in India in the late 1980s, as an underground hippie scene gained international recognition. The genre's high tempo and focus on builds and breaks has fed numerous electronic genres, including modern EDM.

Playlist

Chaiyya Chaiyya
Sukhwinder Singh, Sapna Awasthi
Genre: Filmi

Lahore
Guru Randhawa
Genre: Pop

Dhun
Ravi Shankar
Genre: Hindustani classical

Chaar Bottle Vodka
Yo Yo Honey Singh
Genre: Hip-hop

Krishnanee Begane
Lalgudi Jayaraman
Genre: Carnatic

Mundian To Bach Ke
Panjabi MC
Genre: Bhangra/Hip-hop

Feel
Snow Flakes
Genre: Goa trance

Kabhi Khushi Kabhie Gham
Lata Mangeshkar
Genre: Filmi

Entammede Jimikki Kammal
Vineeth Sreenivasan & Ranjith Unni
Genre: Filmi

Jai Ho
A R Rahman
Genre: Pop

Indonesia

Read List

This Earth of Mankind
by Pramoedya Ananta Toer (1980)
Pramoedya – one of Indonesia's most well-known authors – conceived this story while a jailed political prisoner. It's set in Java during the final years of Dutch rule and explores the injustices endured by Indonesians.

Man Tiger
by Eka Kurniawan (2004, trans 2015)
Longlisted for the 2016 Man Booker International Prize – the only Indonesian writer to ever be nominated – is this story of Margio, a young man who so happens to be a white tiger in human form.

The Land of Five Towers
by Ahmad Fuadi (2009, trans 2011)
Based on a true story, Fuadi's debut novel remains one of the bestselling books in Indonesia. It takes place in a strict Islamic boarding school in East Java and gives fresh insights into religion, nationalism and modern Indonesia.

Saman
by Ayu Utami (1998, trans 2005)
A landmark work for its exploration of sexuality from a middle-class female perspective, Utami's debut novel is set in the Suharto era and tackles taboo subjects.

Harimau! Harimau!
by Mochtar Lubis (1975, trans 1991)
Intended as a metaphor for the blind followers of President Sukarno, '*Tiger!*' tells the story of a group of dammar resin gatherers who follow a spiritual leader into the jungles of Sumatra and find trouble.

Watch List

The Act of Killing
(2012; documentary; directed by Joshua Oppenheimer)
Nominated for an Academy Award for Best Documentary in 2013 is this confrontational and often bizarre series of interviews from the perpetrators accused of the 1965 slaughter of communist sympathisers in Indonesia.

Marlina the Murderer in Four Acts
(2017; drama; directed by Mouly Surya)
Screened at the 2017 Cannes Film Festival, this 'Satay Western' takes places on the island of Sumba and tells the story of two women seeking revenge following a series of wretched events.

The Raid
(2011; action; directed by Gareth Evans)
One of the few Indonesian action films to achieve US box office success, this thriller is set in the slums of Jakarta where local drug lords are set upon by special agents.

Denias, Singing on the Cloud
(2006; drama; directed by John de Rantau)
This beautifully shot film is based on the remote island of Papua. It follows the true, inspiring story of a young boy who struggles against poverty and social injustices.

Leaf on a Pillow
(1998; drama; directed by Garin Nugroho)
Featured at the 1998 Cannes Film Festival, *Leaf on a Pillow* is set in the slums of Yogyakarta, with roles cast to actual street kids over actors. It's a story of hope, perseverance and heartbreak.

Witness daily life in East Java through engaging novel *The Land of Five Towers.*

Channelling Bob Dylan in sound and sensibility, *Guru Oemar Bakrie* is an upbeat, jangly number. It launched Iwan's career and remains one of Indonesia's most famous protest songs, exploring the irony and injustice of how poorly paid teachers are important in society.

Renowned Sudanese composer Gugum Gumbira founded the Jugala Orchestra in the 1960s as a protest against President Sukarno's ban on popular music. The assemblage incorporates traditional West Javanese instruments and performance into its songs.

Playlist

Kuta Rock City
Superman is Dead
Genre: Punk

Kebyar Kebyar
Gombloh
Genre: Rock

Guru Oemar Bakrie
Iwan Fals
Genre: Country/Folk

Bengawan Solo
Oslan Hussein & Teruna Ria
Genre: Pop

Bertamasja
Dara Puspita
Genre: Garage rock

Monumen
The Trees & the Wild
Genre: Indie

Bardin
Jugala Orchestra
Genre: Traditional/Sudanese

Bebas
Iwa K
Genre: Rap

Begadang
Rhoma Irama
Genre: Traditional/Pop

Jarum Neraka
Nicky Astria
Genre: Rock/Pop

Japan

Read List

Norwegian Wood
by Haruki Murakami (1987, trans 1989)
The novel that made Murakami a literary star, this is a coming-of-age story about a young man's exploration of love while dealing with loss, against a backdrop of university days in 1960s Tokyo.

Kitchen
by Banana Yoshimoto (1988, trans 1993)
A haunting tale of two mothers that explores themes of loneliness, transsexuality and hope. *Kitchen* topped the Japan bestseller list for over a year following its release.

Convenience Store Woman
by Sayaka Murata (2016, trans 2018)
An oddball, poignant and heartwarming story of an alienated young convenience store worker existing on the periphery of 'normal' Japanese society, trying to find a new way to fit in.

Snow Country
by Yasunari Kawabata (1948, trans 1957)
Nobel Prize winner Kawabata's masterpiece novel is about a passionate yet doomed love affair between a wealthy dilettante and a country geisha, set in the isolated beauty of western Japan's snowy mountains.

The Pillow Book
by Sei Shōnagon (1002)
A Japanese literature classic, this is a detailed account of the author's life as a lady-in-waiting, offering the reader a fascinating insight into Japanese court life at the height of Heian culture.

Watch List

Seven Samurai
(1954; action; directed by Akira Kurosawa)
This thrilling tale of villagers who hire warriors to protect them against invading bandits is a 3½-hour epic, regarded as one of the most influential films ever made.

Shoplifters
(2018; drama; directed by Hirokazu Kore-eda)
Acclaimed director Kore-eda handles this devastating drama with sensitivity and powerful emotion. It's about a poor, dysfunctional multigenerational family living together on the outskirts of society, trying to survive.

Spirited Away
(2001; anime; directed by Hayao Miyazaki)
Japan's top-grossing film for 20 years (until December 2020), the superb Oscar-winning anime centres on a young girl navigating the spirit world in order to free her parents.

Tokyo Story
(1953; drama; directed by Yasujirō Ozu)
A profound story of a retired couple who make the rare trip to visit their adult children in Tokyo, exploring themes of loss and generational differences.

Jiro Dreams of Sushi
(2011; documentary; directed by David Gelb)
A fascinating profile of the extreme dedication of then-octogenarian sushi master Jiro Ono, owner of the previously Michelin-starred Tokyo restaurant Sukiyabashi Jiro, and his relentless pursuit of perfection.

Tokyo is at the heart of Murakami's *Norwegian Wood* and the film *Tokyo Story*.

Often referred to as the 'Japanese Beatles', Happy End formed in the late 1960s and were influential in bringing the Japanese language to rock. This song featured in Sofia Coppola's film *Lost in Translation*.

A founding member of the pioneering late 1970s Yellow Magic Orchestra band, Sakamoto went on to become a successful solo artist and Oscar-winning film score composer. *Riot in Lagos* is considered an important influence in the development of electronic music as a genre.

Playlist

Hikouki Gumo
Yumi Matsutoya (Yumi Arai)
Genre: Pop

A Day at the Factory
Shonen Knife
Genre: Alt rock

Kaze wo Atsumete
Happy End
Genre: Folk rock

Shock City
Boredoms
Genre: Noise/Alt rock

Farewell
Boris
Genre: Experimental/Noise

Skip
MCpero
Genre: Rap

Riot in Lagos
Ryuichi Sakamoto
Genre: Electronic

PonPonPon
Kyary Pamyu Pamyu
Genre: J-Pop

Dressed in Black
Teengenerate
Genre: Punk

The Barracuda
The 5.6.7.8s
Genre: Garage rock

IN DEPTH

Anime

from JAPAN

Left to right: the onsens of the town of Ginzan in Yamagata, on which the bathhouses of *Spirited Away* are said to be based; Hayao Miyazaki; Totoro and the Catbus from *My Neighbour Totoro.*

Anime is simply the term for Japanese animation created by hand or computer and is closely related to manga (Japanese comics), with many anime films and TV series having been adapted from manga itself. Anime in Japan generates around US$19 billion in revenue annually and has a significant influence on Japanese pop culture, helping give rise to the *otaku* subculture – often referred to as 'geek' culture; generally young people obsessed with pop culture, particularly anime and manga and who like to dress up in cosplay (costumes).

Katsuhiro Ōtomo's 1988 cult classic *Akira*, a cyberpunk film set in a dystopian Tokyo in 2019, is well known around the world, and other anime that gained global recognition include *Mobile Suit Gundam* and *Ghost in the Shell*. But, by far, the most famous and commercially successful anime comes out of the animation studio, Studio Ghibli, with many films by one of its founders, the acclaimed director Hayao Miyazaki. His films include *Nausicaä of the Valley of the Winds* (1984), *My Neighbour Totoro* (1988), *Howl's Moving Castle* (2004), *The Wind Rises* (2013) and *Spirited Away* (2011), which won an Oscar for Best Animated Feature and was Japan's highest-grossing film for almost two decades, until late 2020 when it was topped by the anime film *Demon Slayer*, based on a manga series.

Kazakhstan

Read List

Book of Words
by Abay Qunanbayev (1909)
The most famous work from Kazakhstan's most renowned writer-philosopher-poet sets forth a philosophical path for Kazakhs to follow into the modern era – and remains a revered collection of essays.

A Life At Noon
by Talasbek Asemkulov (2003, trans 2019)
The semi-biographical account of the scion of a family of musicians interprets Kazakh life, landscapes, culture and history through the sounds of the *dombyra* strings plucked by his hands.

Path of Abay
by Mukhtar Auezov (1942, trans 1957)
Based on second-hand accounts of the author's grandfather, a contemporary of Abay Qunanbayev, this work examines the life of the great writer himself and the largely lost customs of turn-of-the-century Kazakhs.

Blood and Sweat
by Abdizhamil Nurpeisov (1961-70, trans 2013)
Whether living under the heel of local strongmen on the rural Kazakh steppe or Soviet authority in far-away Moscow, very little fundamentally changes in the lives of Nurpeisov's long-suffering protagonists.

The Silent Steppe
by Mukhamet Shayakhmetov (2006)
A first-hand account of repression, famine and suffering under the Soviet Union for a family of Kazakhstan's nomadic herders during collectivisation before WWII.

Watch List

The Needle
(1988; drama; directed by Rashid Nugmanov)
One of the first Soviet films to openly confront the social issues of drug addiction and organised crime, *The Needle* is considered a Kazakh classic.

Nomad: The Warrior
(2015; drama; directed by Sergei Bodrov & Ivan Passer)
Kazakh historical fiction hits Hollywood as a descendant of Chinggis (Genghis) Khan seeks to unite the Kazakh tribes for war and defend the Kazakh Khanate.

Nagima
(2013; drama; directed by Zhanna Issabayeva)
Struggling to survive in a world of orphanhood and poverty, an austere life offers little for a group of have-nots on the edges of Kazakh society.

Tale of a Pink Bunny
(2010; drama; directed by Farkhat Sharipov)
Moving away from Kazakh arthouse cinema towards a more mainstream audience, *Pink Bunny* examines the dark sides of Almaty life even as they are viewed through rose-tinted frames.

The Owners
(2014; dark comedy; directed by Adilkhan Yerzhanov)
Originally titled '*Ukkili kamshat*', this film follows a trio of orphaned siblings who return to their ancestral village to reclaim their mother's home. There they find that might (and having connections) counts more than right.

Read *Blood and Sweat* for an understanding of the spartan Kazakh steppe.

The Kazakh national anthem – the lyrics of which were updated in 2006 by then-President Nursultan Nazarbayev – speaks to the bounteous nature and open spaces that largely define Kazakhstan in the Kazakh mindset.

A master of the traditional *kyui* form of Kazakh folk music, Kurmangazy is celebrated as both a childhood prodigy and a father of the art. The Kazakh National Conservatory was named in his honour.

Playlist

Juliya
Batyrkhan Shukenov
Genre: Pop

Tulpar
Galya Bisengalieva
Genre: Classical

My Qazaqstan
Shamshi Kaldayakov
Genre: Patriotic

Ayiptama
Ninety One
Genre: Q-Pop

Daididau
Dimash Kudaibergen
Genre: Folk

Jadua
Jah Khalib
Genre: Rap

Akbai
Kurmangazy Sagyrbaev
Genre: Folk

Er Turan
Turan Ethno-Folk Ensemble
Genre: Folk

Music and Microbes
Nazira
Genre: Electronic

Manmanger
Akhan Seri
Genre: Folk

Kyrgyzstan

Read List

Manas Epic
(995 CE)
Originally an oral tradition and centre of the Kyrgyz origin story, the world's longest epic poem (500,000 lines!) was first recorded and codified by poet Kasym Tynystanov starting in 1926.

Jamilia
by Chingiz Aitmatov (1958, trans 1964)
This novelette by Kyrgyzstan's most famous artistic son follows the scandalous romance of a newly-wed village bride left behind during WWII and an injured soldier recently returned.

When the Edelweiss Flowers Flourish
by Begenas Sartov (1969, trans 2012)
Traditional pastoral lifestyles and the inherited wisdom of homeopathic remedies combine with an unexpected extraterrestrial theme in this Soviet-era work by Kyrgyzstan's first science fiction author.

Blood Ties and the Native Son: Poetics of Patronage in Kyrgyzstan
by Aksana Ismailbekova (2017)
This fascinating work offers insight into how centuries of clan and kinship politics continue to inform Kyrgyz politics and democratisation in the modern era.

Broad River
by Tugelbay Sydykbekov (1938)
Regarded as the first true Kyrgyz novel, *Broad River* launched a writing career that won Sydykbekov the Stalin Prize and recognition as a hero of the Kyrgyz Republic.

Watch List

Boz Salkyn
(2007; drama; directed by Ernest Abdyjaparov)
City girl Asema is kidnapped as a bride-to-be in a case of mistaken identity, but grows to love her simple life in the Kyrgyz mountains.

Kurmanjan Datka
(2014; action; directed by Sadyk Sher-Niyaz)
History lives in this fictionalised story of the 'Queen of the Alay', a revered historical figure whose diplomacy held off the advance of Imperial Russia.

Metal Bread
(2014; documentary; directed by Chingiz Narynov)
Life is hard and unchanging in the Soviet-era mining town of Mailuu-Suu, where resident Tatyana rifles through broken glass to earn enough to buy her daily bread.

Salam, New York!
(2013; comedy; directed by Ruslan Akun)
An aspiring but unprepared young Kyrgyz man leaves for the bright lights and bright future of New York City, with endless challenges along the way.

Beshkempir the Adopted Son
(1998; drama; directed by Aktan Abdykalykov)
Explore the unusual custom of offering children from large families for adoption to childless couples in this coming-of-age story that received international acclaim. The star of the film was the director's son, now a director himself.

Rural village life features in the Kyrgyz film *Boz Salkyn* and the book *When the Edelweiss Flowers Flourish*.

Kara Jorgo is a centuries-old performance, but the craze for hearing it anywhere and everywhere is a wholly modern phenomenon. Evoking the image of whipping a horse while hunting for wild prey, you'll eventually get dragged into dancing this if you visit Kyrgyzstan.

Calling for rights and equality for women, this track by Zere caused quite a stir in Kyrgyzstan's still very patriarchal society when it was released in 2018. It generated both an upswell of community support and quite a few death threats for the artist.

Playlist

Jarashat
Bayastan
Genre: Hip-hop

Kara Jorgo
Asel Kasmanova
Genre: Folk

Emotional Tango
Askat Jetigen uulu
Genre: Classical

Apakem
Gulnur Satylganova
Genre: Pop

Ala-Too
Salamat Sadykova
Genre: Folk

Mash Botoy
Ordo Sakhna
Genre: Folk

Kyz
Zere
Genre: Pop

Diko
Ulukmanapo
Genre: Rap

Sen Zhalgyz
Kanykei
Genre: Pop

Zhanym
Mirbek Atabekov
Genre: Pop

Malaysia

Read List

The Kampung Boy
by Lat (1979)
Malaysia's beloved cartoonist mines his own life story in this gently humorous graphic novel about Mat, a young boy growing up in a jungle village in the 1950s.

The Harmony of Silk Factory
by Tash Aw (2005)
Aw's award-winning debut novel, set in British colonial Malaya of the interwar years, is a torrid tale of love and betrayal told from three perspectives.

The Gift of Rain
by Tan Twan Eng (2007)
In this Man Booker Prize-nominated novel, fraternal, familial and national loyalties are tested on the island of Penang, before and during WWII.

Peninsula: A Story of Malaysia
by Rehman Rashid (2016)
Rashid's memoir is an engaging overview of the country's recent political and social events, and how they are rooted in the multicultural strands of Malaysia's history.

Once We Were There
by Bernice Chauly (2017)
Sex, drugs, street protests, a doomed marriage and a stolen baby are the key elements of this gritty novel set in Kuala Lumpur in the politically turbulent 1990s.

Watch List

Penarek Becha (The Trishaw Man)
(1955; drama; directed by P Ramlee)
Ramlee, an icon of Malay cinema, was just 26 when he wrote, directed and starred in this Singapore-set tale of a rich man's daughter falling for a poor trishaw driver.

Sepet
(2005; drama; directed by Yasmin Ahmad)
Despite its controversial subject matter, Ahmad's rom-com about the blossoming love affair between a Chinese boy and a Muslim Malay girl was a local hit.

Bunohan
(2012; drama; directed by Dain Said)
A languid mood hangs over this arthouse thriller about a kickboxer on the run and troubled family relations – it gains much from its setting on the swampy Kelantan coast.

M for Malaysia
(2019; documentary; directed by Dian Lee & Ineza Roussile)
Go behind the scenes of 2018's landmark general election; co-director Roussile is the granddaughter of Mahathir Mohamad, the two time PM who features heavily in the footage.

The Garden of Evening Mists
(2020; drama; directed by Tom Shu-Yu Lin)
A Malaysian woman is haunted by the death of her sister and a relationship with an enigmatic Japanese gardener in this historical romance based on Tan Twan Eng's award-winning novel.

Kuala Lumpur's skyline has changed since the city featured in *Once We Were There*.

The year he died in 1973, aged just 44, the Malaysian silver screen legend P Ramlee composed *Tears of Kuala Lumpur* for his wife, the singer Saloma. In her 1960s heyday, she was known as the 'Marilyn Monroe of Asia'.

With a soulful voice to match her sultry looks, Yuna saw her songs go viral when she uploaded them to Myspace. The poster girl of Malaysian 'hijabsters' now records in LA with the likes of Usher and Pharrell Williams.

Playlist

Kumpulan Muzik Gambus
Fadzil Ahmad Ensemble
Genre: Folk

Bukan Cinta Biasa
Siti Nurhaliza
Genre: Pop

Air Mata di Kuala Lumpur
Saloma
Genre: Ballad

Hijau
Zainal Abidin
Genre: Pop

Angin Kencang
Noh Salleh
Genre: Pop

Jelita
Kyoto Protocol
Genre: Indie rock

Dance Like Nobody's Watching
Yuna
Genre: R&B

No Christmas For Me
Zee Avi
Genre: Indie pop

Alhamdulillah
Too Phat ft Yasin
Genre: Hip-hop

Togok
NJWA ft CEE & Gangsapura
Genre: Folk electronica

Mongolia

Read List

My Native Land
by Dashdorjiin Natsagdorj (c1930s)
One of the founders of modern Mongolian literature, Dashdorjiin Natsagdorj's most famous work is this 13 stanza poem that sings praise to his homeland's natural landscapes and moral character.

The Blue Sky
by Galsan Tschinag (1994, trans 2006)
A semi-autobiographical novel based on Tschinag's experience as a child growing up in the Altai mountains while trying to maintain his traditional nomadic Tuvan way of life in the face of a changing world.

Ghengis Khan and the Making of the Modern World
by Jack Weatherford (2005)
Groundbreaking in its approach, this bestselling historical account on the Mongol empire reveals just how influential and far-reaching the legacy of Chinggis Khaan was in shaping Western civilisation.

My Mongolian World: From Onon Bridge to Cambridge
by Urgunge Onon (2006)
Offering a fascinating insight into Mongolian culture is this memoir of a respected scholar who draws upon his own experiences, from shamanistic encounters and wolf hunting to dealing with royalty and being kidnapped.

The Secret History of the Mongols
Anonymous (13th century)
Legends, oral and written history of the Mongol people and its rise to power. It was considered 'secret' due to only members of the imperial family being permitted to read it.

Watch List

The Story of the Weeping Camel
(2003; docudrama; directed by Byambasuren Davaa & Luigi Falorni)
Both heartbreaking and joyous, this docudrama follows a family of herders in the Gobi Desert and their efforts to restore the bond between a baby camel and its mother.

Mongol
(2007; historical drama; directed by Sergei Bodrov)
This epic blockbuster is a dramatic depiction of the rise of Chinggis Khaan, featuring stunning landscapes and a storyline mixing romance, drama and battle scenes.

Tracking the White Reindeer
(2009; docudrama; directed by Hamid Sardar)
Taking place in the snow-covered steppes of northern Mongolia, this beautifully shot docudrama tells the story of a nomadic Tsaatan boy's journey to recover an escaped reindeer.

The Eagle Huntress
(2016; docudrama; directed by Otto Bell)
A 13-year-old Kazah girl goes on a quest to become the first female champion in Bayan-Ölgii's eagle-hunting competition in far western Mongolia.

Khadak
(2006; drama; directed by Peter Brosens & Jessica Hope Woodworth)
Set on the Mongolia steppes in the height of winter, this story is about an epileptic nomadic herder who, after being forced to relocate, loses his way of life before another path opens.

This enormous statue of Mongol warlord Chinggis Khaan is just east of the capital city Ulaanbaatar.

As ethereal as it is haunting, this transcendental song by Namjilyn Norovbanzad is on a different plain. Known as the queen of traditional Mongolian long songs, Norovbanza was voted 'singer of the century' by the Mongolian people in the year 2000.

The Hu aren't the first band to meld traditional Mongolian instruments, throat singing and ancient war cries with rocking guitar riffs, but this epic track makes them the first local band to reach number one on the Billboard Hard Rock Digital Song Sales chart.

Playlist

Zuud Noirondoo
The Lemons
Genre: Indie

Uyhan Zambuu Tiviin Naran
Namjilyn Norovbanzad
Genre: Traditional

Khukh Tolboton
Altan Urag
Genre: Traditional/Rock

Talin Salhi
Egschiglen
Genre: Traditional/Rock

Mongol
Khusugtun
Genre: Folk

Banquet (Mongolian version)
Magnolian
Genre: Indie folk

Wolf Totem
The Hu
Genre: Traditional/Rock

Enkh mendiin bayar
Myagmarsürengiin Dorjdagva
Genre: Traditional

Hood
Vanquish ft TG, Gee, Desant
Genre: Rap

Tsas Narand Uyarna
Nominjin
Genre: Pop

Myanmar

Read List

From the Land of Green Ghosts
by Pascal Khoo Thwe (2002)
A beautifully written memoir that's also a thrilling page-turner as Khoo Thwe recounts his journey from a remote tribal village in Myanmar to becoming a student at Cambridge University.

The River of Lost Footsteps
by Thant Myint-U (2006)
The grandson of U Thant, former secretary-general of the United Nations, elegantly interweaves Myanmar's history with that of his family, illuminating the key events that have shaped the modern-day nation.

Smile as They Bow
by Nu Nu Yi (1994, trans 2008)
Nominated for the Man Asian Literary Prize, this taboo-busting novel is set against the background of the Taungbyone nat spirit festival, with an ageing gay transvestite medium as its central character.

Golden Parasol: A Daughter's Memoir of Burma
by Wendy Law-Yone (2013)
Law-Yone's father was an exiled Burmese newspaper editor. Her memoir explores hopes for democracy and how they were dashed by the military coup of 1962.

Miss Burma
by Charmaine Craig (2018)
Inspired by the lives of Craig's grandparents and mother, this novel is about a 1950s beauty queen quitting Rangoon high society to lead a guerrilla army fighting for an independent Karen state.

Watch List

The Lady
(2011; drama; directed by Luc Besson)
Michelle Yeoh is well cast as Aung San Suu Kyi in this biopic about Myanmar's most famous political activist and long-time prisoner of conscience.

Kayan Beauties
(2012; drama; directed by Aung Ko Latt)
Realistic and gritty drama in which three tribal Kayan women attempt to rescue a young Kayan girl from enslavement by human traffickers.

The Monk
(2014; drama; directed by The Maw Naing & Maw Naing Aung)
The teenager protagonist in this film has to decide whether to remain a monk in this delicate character study by poet and artist The Maw Naing.

Twilight over Burma: My Life as a Shan Princess
(2015; drama; directed by Sabine Derflinger)
True-life story of Inge Eberhard (now Sargent) who battled to free her husband, a Shan prince, following the military coup of 1962 in Myanmar.

Golden Kingdom
(2015; drama; directed by Brian Perkins)
Filmed entirely on location in Myanmar, this drama is about four young monks living in a remote monastery.

Life in a Myanmar monastery is explored in the movie *Golden Kingdom*.

Written in the 1980s, *Way Twar Tae A Khar (When We Drift Apart)* is one of the most popular songs from the man considered to be the grandfather of Burmese pop. Thin has been greatly mourned in Myanmar since his premature death in 2004 aged 41.

This 2020 single is by the vanguards of the Burma's punk and indie-rock scene, who've been jamming since 2004. Yangon-based Side Effect have played many international gigs, including the 2014 SXSW festival in Austin.

Playlist

Shwe Ka Nyar
Kyaw Kyaw Naing
Genre: Traditional

Pyan Sone Si Kwint
May La Than Zin
Genre: Pop

Way Twar Tae A Khar
Htoo Eain Thin
Genre: Pop

Hmone Shwe Yi
Tin Tin Mya
Genre: Folk

Yone Kyi Yar
Lay Phyu
Genre: Rock

War
Phyu Phyu Kyaw Thein
Genre: Pop

Chit Oo May
Doe Lone
Genre: Rock

Red Light
Sai Sai Kham Leng x Park Bom
Genre: Hip-hop

Aye Say Bah
Side Effect
Genre: Indie rock

Starry Night
Big Bag
Genre: Indie rock

Nepal

Read List

Arresting God in Kathmandu
by Samrat Upadhyay (2001)
A winning collection of short stories that paints a picture of a society caught between tradition and modernity, struggling to reconcile newfound freedoms with the constraints of faith and family.

The Tutor of History
by Manjushree Thapa (2001)
Nepal's tricky transition from monarchy to democracy is the backdrop to this novel of party politics in a country village, with corruption almost a character in its own right.

Mountains Painted with Turmeric
by Lil Bahadur Chettri (1957, trans 2008)
A Nepali favourite, this novel of village life in far eastern Nepal brings evocatively to life the *dukha* (suffering) of the peasant farmers at the bottom of Nepal's social order.

Shirishko Phool
by Parijat (1964, trans 1972)
In 'Blue Mimosa', a Gurkha veteran overcome by feelings of *shunya* (emptiness) becomes infatuated with the sister of a drinking buddy, forcing him to re-examine acts of violence from his past.

Muna Madan
by Laxmi Prasad Devkota (1936)
This classic poem is one of the bestselling works of Nepali literature, a tale of a new marriage strained by distance and doomed to tragedy, unfolding on both sides of the Nepal-Tibet border.

Watch List

Mukundo (Mask of Desire)
(2000; drama; directed by Tsering Rhitar Sherpa)
In this taboo-breaking drama, an expectant mother desperate to bear a son makes a pact with a holy man and enters a world of possession, exorcism and jealousy.

Kagbeni
(2008; horror; directed by Bhusan Dahal)
Nepal's first digital movie shifts WW Jacobs' vintage horror tale *The Monkey's Paw* to Nepal, as a wish-fulfilling fetish lures a group of friends towards disaster.

Chino
(1991; drama; directed by Tulsi Ghimire)
This Nepali smash hit overflows with villainy and revenge in the Bollywood tradition, complete with rousing song and dance numbers, choreographed fight scenes and menacing moustaches.

Even When I Fall
(2017; documentary; directed by Sky Neal & Kate McLarnon)
A British-Nepali collaboration, telling the remarkable true story of the young women who escaped from human trafficking in India and founded Nepal's first circus company.

Chhakka Panja
(2016; comedy; directed by Deepa Shree Niraula)
Spawning two sequels, this social satire touches on key Nepali themes – marriage, lost love, unemployment and the lure of work overseas.

Some of Samrat Upadhyay's short stories are set in the Nepalese capital Kathmandu.

A pop career might not seem an obvious move for a Buddhist nun, but Ani Choying Drolma was pivotal in bringing Tibetan music to a mainstream audience, touring with American guitarist Steve Tibbetts and even recording a session for MTV.

What do old rock stars do when they retire? In the case of Deepak Rana, lead guitarist with Nepathya, the answer was become a helicopter pilot, joining Iron Maiden's Bruce Dickinson and Gary Numan in the ranks of famous flying rockers.

Playlist

Mayalule
1974 AD
Genre: Folk rock

Aath Din
The Shadows
Genre: Rock

Chö
Ani Choying Drolma
Genre: Buddhist chanting

Timro Jasto Mutu
Narayan Gopal
Genre: Folk

Bhool Ma Bhulyo
Robin N Looza
Genre: Rock

Maya
The Uglyz
Genre: Rock

Yo Jindagani
Nepathya
Genre: Folk rock

Ma Maunta Ma
Om Bikram Bista
Genre: Nepali pop

Disconnect
Underside
Genre: Metal

Phariya Lyaaidiyechan
Navneet Aditya Waiba
Genre: Folk

Pakistan

Read List

A Case of Exploding Mangoes
by Mohammed Hanif (2008)
A darkly comic imagining of the backstory behind the mysterious death of former Pakistani president General Zia ul-Haq, delving into the complex relationship between the government and military in Pakistan.

Asrar-i-Khudi
by Muhammad Iqbal (1915, trans 1920)
The first book by Pakistan's national poet Muhammad Iqbal, '*The Secrets of the Self*' explores the philosophy of faith and the quest to reveal the divine spark that resides inside every being.

The Reluctant Fundamentalist
by Mohsin Hamid (2007)
A conversation in a cafe in Lahore evolves into a parable about America's role in driving fundamentalism, as a Pakistani professor tells his life story to an American stranger.

The Crow Eaters
by Bapsi Sidhwa (1978)
A brilliantly irreverent comedy about a Parsi family forging a new life in British-governed Lahore, against a backdrop of social climbing, colonialism and family politics.

Mottled Dawn
by Saadat Hasan Manto (1997)
Fifty short stories, translated from Urdu, that weave a powerful narrative about the tragedies that befell Hindus and Muslims as Partition divided India and Pakistan in 1947.

Watch List

Cake
(2018; drama; directed by Asim Abbasi)
A smartly comic family drama, spanning Karachi, interior Sindh and London, with strong female leads and wry observations of family struggles.

Manto
(2015; biopic; directed by Sarmad Khoosat)
A life story of the respected Urdu writer Saadat Hasan Manto, his relationship with singer Noor Jehan, and the frequent attempts to prosecute him for obscenity.

Khuda Kay Liye
(2007; thriller; directed by Shoaib Mansoor)
Two Pakistani singers find their lives turned upside down by the events of 9/11, in a film that opened doors between India and Pakistan,

Maula Jatt
(1979; action; directed by Younis Malik)
Rival clans, dishonour, revenge, bombastic machismo and theatrical fight scenes – *Maula Jatt* set the template for Lollywood (the Pakistani film industry) for a generation.

Anarkali
(1958; drama; directed by Anwar Kamal)
A slave girl falls in love with a prince who rises to become a Mughal emperor. Expect lavish costumes, flamboyant songs and limited historical accuracy!

Coming from a family of traditional *qawwali* (devotional hymn) singers, Nusrat Fateh Ali Khan introduced the world to Pakistani music. He played at WOMAD, collaborated with Peter Gabriel and Eddie Vedder, and wrote music for a string of Pakistani and Bollywood films.

The songbird of Pakistani *filmi* (soundtrack) music, Noor Jehan was astonishingly prolific, recording some 20,000 songs in a performing career spanning six decades. She also found time to act in over 40 movies and became Pakistan's first female film director.

Lahore lends its name to the Pakistani movie industry, Lollywood.

Playlist

Desi Dab
Young Desi
Genre: Rap

Democracy is the Best Revenge
Dead Bhuttos
Genre: Punk/Hardcore

Haq Ali Ali
Nusrat Fateh Ali Khan
Genre: Qawwali

Sayonee
Junoon
Genre: Sufi rock

Kitni Sadiyaan
Mizraab
Genre: Prog rock

Dil Dil Pakistan
Vital Signs
Genre: Pop/Rock

Dang Pyar Da Seene Tei Kha Kei
Noor Jehan
Genre: Filmi

Duur
Strings
Genre: Rock

Gulon mein rang bharay
Mehdi Hassan
Genre: Ghazal

Dupatta
Hadiqa Kiani
Genre: Pop

Philippines

Read List

Po-On
by F Sionil José (1984)
'*Dusk*' is the start of a five-volume historical epic, tracing the troubled exile of a tenant farming family to the village of Rosales, set against a backdrop of colonial oppression and revolution.

Smaller and Smaller Circles
by F H Batacan (2002)
Crime drama, Philippines-style: a duo of Jesuit monks turn to forensic science when the corrupt authorities fail to stop a spate of murders in a *barangay* (sub-district) of Metro Manila.

May Day Eve
by Nick Joaquin (1947)
A strange, often profound exploration of a troubled relationship. The magical realism employed in this short story speaks volumes about the conceits and insecurities of the Filipino elite under Spanish rule.

Noli Me Tángere
by José Rizal (1887, trans 1912)
The key work by one of the Philippines' national heroes, exploring the plight of the islands under Spanish rule. It's full of intrigue, betrayal, class struggle and racial politics.

Florante at Laura
by Francisco Balagtas (1869)
A fantastical reimaging of events borrowed from European classical history, executed in the traditional Filipino Awit poetic form, with four lines per stanza and 12 syllables per line.

Watch List

Himala
(1982; drama; directed by Ishmael Bernal)
In this haunting film, a biblical vision spawns a religious movement, masking a terrible crime that eventually engulfs all of its protagonists.

Kid Kulafu
(2015; sports; directed by Paul Soriano)
The Philippines' answer to *Rocky*, this ambitious biopic explores the tough childhood of boxer Manny Pacquiao, and his journey from destitution to superstardom.

Ma' Rosa
(2016; drama; directed by Brilliante Mendoza)
Showing the growing maturity of the Pinoy film industry, *Ma' Rosa* follows a convenience-shop owner pulled by poverty into a dangerous world of drug dealing and police corruption.

Toto
(2015; comedy; directed by John Paul Su)
A nuanced, funny look at the Filipino obsession with the American dream, and the lengths many Filipinos go to in order to work overseas. It's told through the eyes of a hotel room attendant with big dreams.

Maynila, sa mga Kuko ng Liwanag
(1976; drama; directed by Lino Brocka)
Perhaps the greatest work by subversive director Lino Brocka, '*Manila in the Claws of Light*' is a gritty tale of disillusionment about a young man who seeks a new life in Manila, only to be confronted by the past.

Surviving in the metropolis of Manila is the topic of *Maynila, sa mga Kuko ng Liwanag*.

Outwardly, Eraserheads' hit *Ang Huling El Bimbo* is a ballad of unfulfilled love (with a macabre twist). Yet it even spawned a stage musical, with the song providing the storyline linking a back catalogue of Eraserheads hits.

Freddie Aguilar's autobiographical 1978 track *Anak* – the tale of a wayward son and the parents he left behind – was not just a smash in the Philippines. The song raced up the charts in Japan, Hong Kong, Malaysia, the US, Europe and Angola.

Playlist

Basang-Basa Sa Ulan
Aegis
Genre: Pop/Rock

Himig Natin
Juan De La Cruz Band
Genre: Pinoy rock

Ang Huling El Bimbo
Eraserheads
Genre: Indie rock

Kahit Kailan
South Border
Genre: R&B

Narda
Kamikazee
Genre: Indie rock

Sundo
Imago
Genre: Pop/Rock

Anak
Freddie Aguilar
Genre: Folk

Before I Let You Go
Freestyle
Genre: R&B

Tala
Sarah Geronimo
Genre: Pop

Sana Ngayong Pasko
Jimmy Borja
Genre: Easy Listening

Singapore

Read List

How We Disappeared
by Jing Jing Lee (2000)
In this evocative, page-turning novel, a Singaporean widow confronts the brutal events of the city-state's occupation by the Japanese during WWII that altered the course of her life forever.

Singapore Story
by Lee Kuan Yew (1998)
To get the official story on Singapore's transformation from poor, Southeast Asian backwater to economic powerhouse, go straight to the man who masterminded it – the founding Prime Minister of Singapore.

Little Ironies: Short Stories of Singapore
by Catherine Lim (1978)
The doyenne of Singapore fiction has published numerous short-story collections and novels. Spotlighting ordinary people at their best and worst in a rapidly modernising Singapore, this is Lim's first collection.

Inheritance
by Balli Kaur Jaswal (2013)
A nation's coming-of-age story, seen through the lens of a traditional Punjabi family. It gradually unravels over two decades as Singapore's political, social and cultural landscapes change.

Ministry of Moral Panic
by Amanda Lee Koh (2013)
Tackling issues of gender, sexuality, race, social inequality and more, Koh's debut collection of quirky short stories are a reflection on a changing culture.

Watch List

Crazy Rich Asians
(2018; romantic drama; directed by Jon M Chu)
First came the bestselling novel by Kevin Kwan, then came the Hollywood hit that sees a Chinese-American academic navigate the world of Singapore's uber-rich while visiting the island-state with her boyfriend, the heir to a real-estate dynasty.

Money No Enough
(1998; comedy; directed by T L Tay)
In a turning point for the nation's film industry, the box office success of this movie about three Singaporeans facing financial difficulties proved that Singaporean films could be economically viable.

A Yellow Bird
(2016; drama; directed by K Rajagopal)
This story of a Singaporean-Indian ex-convict seeking redemption shows a side of Singapore beyond the glitz many equate with the island state.

To Singapore, With Love
(2013; documentary; directed by Pin Pin Tan)
Banned in Singapore, this film comprises interviews with nine political exiles, all over 60, who were previously student leaders, trade unionists or members of Singapore's communist party.

15
(2003; comedy-drama; directed by Royston Tan)
The stark portrayal of five Chinese-Singaporean youths abandoned by the system (with real-life triad members as actors) has earned cult status in Singapore.

The *Singapore Story* of the city's futuristic metamorphosis is told by Lee Kuan Yew.

One of the top bands in the '60s rock scene, The Quests' instrumental, *Shanty*, was the first song by a local band to top the Singapore charts. It knocked off The Beatles' *I Should Have Known Better.*

A highlight of the National Day Parade for many Singaporeans is the cheesy theme song created for it each year. Written by legendary singer-songwriter Dick Lee, the 1998 National Day song *Home* was particularly popular.

Playlist

I Wouldn't Know Any Better Than You
Gentle Bones
Genre: Pop

You're The Boy
Shirley Nair & The Silver Strings
Genre: Pop

Shanty
The Quests
Genre: Pop/Rock

So Happy
The Oddfellows
Genre: Indie

Usah Lepaskan
Taufik Batisah
Genre: Pop

Xi Shui Chang Liu
Liang Wern Fook
Genre: Xinyao (Singapore folk)

Home
Kit Chan
Genre: Patriotic pop

Not Tonight
JJ Lin
Genre: Electropop

Better
Sam Rui
Genre: R&B

Rasa Sayang
Dick Lee
Genre: Pop

South Korea

Read List

I'll Be Right There
by Shin Kyung-sook (2010, trans 2014)
Taking place in Seoul in the 1980s to the backdrop of pro-democracy protests, this evocative, award-winning novel is about a young women's journey and her dealings with tragic memories that have resurfaced.

Three Generations
by Yom Sang-seop (translated 2005)
This classic novel written by one of Korea's most famous authors was originally published as a newspaper serialisation in the 1930s. It focuses on the hardships of a middle-class family during Japanese occupied Seoul.

The Vegetarian
by Han Kang (2007)
The story of a young woman who suddenly decides to stop eating meat following a vivid series of dreams, and the adverse effects this has upon her relationship with her family.

I Have the Right to Destroy Myself
by Kim Young-ha (1996, trans 2007)
Written by one of the leading voices among contemporary Korean writers is this dark and existentialist novel set in Seoul about two brothers who fall for the same woman.

Who Ate up All the Shinga
by Park Wan-suh (1992, trans 2009)
A semi-autobiographical account of a family torn apart by the Korean War. The novel is both captivating and touching, written in effective, straightforward prose.

Watch List

Parasite
(2019; black comedy thriller; directed by Bong Joon-ho)
Cleaning up in the 2020 Academy Awards – including Best Picture and Best Director – this funny, original and disturbing film is about a wealthy family having their lives infiltrated by a scheming family of lower-class fraudsters.

Chihwaseon
(2002; drama; directed by Im Kwon-taek)
This movie about esteemed 19th-century painter Jang Seung-up is by one of South Korea's most prolific filmmakers. It took the award for Best Director at Cannes in 2002.

Pietà
(2012; thriller; directed by Kim Ki-duk)
A Golden Lion winner at Venice, *Pietà* is a disturbing and powerful film about a ruthless loan shark and a middle-aged woman who claims to be his mother.

On the Beach at Night Alone
(2017; drama; directed by Hong Sang-soo)
Written, directed and produced by indie film maker Hong Sang-soo, this story about an actress embroiled in a scandalous extra-marital affair takes place in both Hamburg and the seaside city of Gangneung.

Train to Busan
(2016; action-horror; directed by Yeon Sang-ho)
A blockbuster apocalyptic zombie horror about a virus outbreak across South Korea that sees train passengers battling a never-ending parade of flesh-eating zombies.

A neighbourhood of Busan, not currently plagued by *Train to Bustan's* flesh-eating zombies.

The smash hit that turned K-pop into an overnight sensation, *Gangnam Style* is as catchy as it is entertaining. All about sending up the ritzy lifestyles of Seoul's socialites, its music video – helped by Psy's quirky dance moves – went viral and (for a time) achieved the most ever likes on YouTube.

A scintillating fusion of traditional *haegum* (Korean stringed instrument) and power chords, *Time of Extinction* is an epic track, and one that turned Korea's past into the new sound.

Playlist

Boombayah
Blackpink
Genre: K-Pop

Love Song
Danpyunsun and the Sailors
Genre: Fusion/Pop

Gangnam Style
Psy
Genre: K-pop

Dynamite
BTS
Genre: K-pop

Gangwon-do Arirang
Song Sohee
Genre: Traditional folk

16-22
Dead Buttons
Genre: Indie

Time of Extinction
Jambinai
Genre: Indie/Traditional

Nan Arayo
Seo Taiji and Boys
Genre: Rap

Heeya
Boohwal
Genre: Rock

Idol
Jaurim
Genre: Indie

IN DEPTH

K-Pop

from SOUTH KOREA

Left to right: BTS, one of the biggest pop bands in the world in the 2020s; girl group Blackpink performing in Seoul; a Psy-inspired sculpture in Seoul's Gangnam neighbourhood .

K-pop, a national obsession, is as much a cultural phenomenon as it is a genre of music. From subculture beginnings to its mainstream craze, it's an industry that's grown exponentially since the 1990s – first sweeping across South Korea before making its move globally as a major player.

So what is exactly K-Pop? Essentially it's a heavy stylised and upbeat form of urban Korean pop music. All about idol culture, acts are manufactured boy/girl bands, each member marketed individually as a star or sex symbol in their own right. Songs are catchy and take on a fusion of anything from R&B, hip hop, rock and electronica, and accompanied by choreographed dance moves. It's all then showcased in high-budget, over-produced video clips.

K-Pop is an integral aspect to Hallyu, aka the Korean Wave, which was the sudden interest in South Korean pop culture (film, TV and fashion) spreading across Asia in the 1980s. And then along came the internet, and YouTube and a little song called *Gangnam Style*, and from that moment everything changed. K-pop turned into an overnight global sensation, and paved the way for bands like BTS and Blackpink who became even bigger, selling out stadiums worldwide and proving that K-Pop is more than a fad – it's here to stay.

Sri Lanka

Read List

Anil's Ghost
by Michael Ondaatje (2000)
A haunting novel about the quest to find a man's identity amid the turmoils of the civil war in Sri Lanka. Michael Ondaatje is a Booker Prize-winning author for his novel *The English Patient*.

Island of a Thousand Mirrors
by Nayomi Munaweera (2012)
A powerful literary debut by Sinhalese author Munaweera, this coming-of-age story of two very different women whose fate is woven together won the 2013 Commonwealth Book Prize for Asia.

On Sal Mal Lane
by Ru Freeman (2013)
Ru Freeman's deeply moving novel is a beautiful story about a community of families living in a lane in Colombo and the complexities of their society amid a looming war.

Cinnamon Gardens
by Shyam Selvadurai (1998)
This thoroughly rewarding and evocative novel set in 1920s Ceylon delves into the complex social and political issues hidden behind the polished façade of upper-class society.

Monkfish Moon
by Romesh Gunesekera (1992)
This expertly and beautifully written collection of nine short stories paints a vivid and diverse picture of contemporary Sri Lanka against a backdrop of natural beauty and ethnic conflict.

Watch List

Mansion by the Lake
(2002; drama; directed by Lester James Peries)
Considered the father of Sri Lankan cinema, Lester James Peries' *Mansion by the Lake*, originally titled '*Wekande Walauwa*' is set in 1980s rural Sri Lanka and is loosely based on Anton Chekov's *The Cherry Orchard*.

Death on a Full Moon Day
(1997; drama; directed by Prasanna Vithanage)
Against a backdrop of the struggles of everyday life in rural Sri Lanka, this highly acclaimed film centres on an old man's refusal to accept government compensation after his son's death in the civil war.

The Forsaken Land
(2005; drama; directed by Vimukthi Jayasundara)
Winner of the Caméra d'Or at the 2005 Cannes Film Festival, this is a disturbing, affecting and visually striking film focusing on the casualties of war.

The Road from Elephant Pass
(2008; action; directed by Chandran Rutnam)
Based on the novel by Nihal de Silva, this war film tells the story of the battle for Elephant Road, a strip of land linking the Jaffna Peninsula with the rest of Sri Lanka.

Karma
(2010; drama; directed by Prasanna Jayakody)
Karma is an emotionally turbulent story of a young man dealing with the guilt of his mother's death, and the complex relationships of three interconnected lives.

Life moves pretty fast in Colombo, setting for *On Sal Mal Lane* by Ru Freeman.

This song skyrocketed British-born rapper and activist M.I.A (Maya Arulpragasam), daughter of a Tamil revolutionary leader, to global fame. It is said to be a satirical response to repressive bureaucracy and was nominated for a Grammy for Record of the Year in 2009.

Clarence Wijewardena is one of Sri Lanka's most respected musicians and considered the father of Sinhalese pop. Along with Annesley Malewana, he formed The Moonstones and later The Super Golden Chimes. He wrote and composed this hit song in the mid-70s.

Playlist

Gum Gum Gum (One Note Song)
Sunil Shantha
Genre: Sinhala folk

Ran Dahadiya Bindu Bindu
W D Amaradeva
Genre: Sinhala folk

Paper Planes
M.I.A
Genre: Hip-hop

I Don't Know Why
The Gypsies
Genre: Baila

Bambara Pahasa
Rookantha Gunathilake
Genre: Pop

Rock n Roll is My Anarchy
Paranoid Earthling
Genre: Rock

Kanda Surinduni
The Super Golden Chimes
Genre: Sinhala pop

Andura
Stigmata
Genre: Metal

Me Geeya Mama Gayanne
Three Sisters
Genre: Pop

Taiwan

Read List

A Thousand Moons on a Thousand Rivers
by Hsiao Li-hung (1980, trans 2000)
Set in the 1970s in Budai, a coastal town in southern Taiwan where the author grew up, this novel blends romance, Buddhist teachings and local traditions.

Barbarian at the Gate: From the American Suburbs to the Taiwanese Army
by T C Locke (2003, trans 2014)
After the US-born author – now known as T C Lin – renounced his American citizenship and became a citizen of Taiwan, he was required to serve two years in the Taiwanese military. This memoir details his experience.

Notes of a Crocodile
by Qiu Miaojin (1994, trans 2017)
This novel about queer youth in 1980s Taiwan was published posthumously after the author, frequently regarded as one of the country's predominant lesbian authors, committed suicide at age 26.

Ghost Month
by Ed Lin (2014)
A food vendor investigates the murder of his former high-school girlfriend in the first of a series of mystery novels that the Taiwanese-American author set in Taipei's delicious night markets.

The Butcher's Wife
by Li Ang (1983, trans 1986)
This controversial novel about a young woman married to an abusive pig butcher is considered a Taiwanese feministic classic, exploring themes of sexuality and gender politics.

Watch List

Crouching Tiger Hidden Dragon
(2000; action-adventure; directed by Ang Lee)
The Taiwanese-American director's Oscar-winning *wuxia* (martial arts) epic features elaborately choreographed fight scenes and an ill-fated love story, set amid the striking landscapes of ancient China.

Eat Drink Man Woman
(1994; comedy-drama; directed by Ang Lee)
One of Ang Lee's trio of films about Taiwanese family life, this movie follows a chef who lives in Taipei with his three grown daughters, who dissect their love lives over weekly Sunday meals.

On Happiness Road
(2017; drama; directed by Sung Hsin-Yin)
Based on the director's own upbringing in Taiwan in the 1970s and '80s, this animated movie – her first feature film – also highlights political issues of the era.

A Sun
(2019; drama; directed by Mong-hong Chung)
In this contemporary drama of two brothers, the older is the family's star and his younger sibling is deemed a problem child, until circumstances cause the family dynamic to shift.

Yi Yi
(2000; drama; directed by Edward Yang)
Hailed as one of the pioneering filmmakers of the Taiwanese New Wave in the 1980s and '90s, Yang wrote and directed this sprawling story that follows three generations of a Taiwanese family.

Taipei's aromatic night food markets are the setting for *Ghost Month* by Ed Lin.

Singing in Puyuma, her traditional language, Samingad is among Taiwan's best-known Indigenous singer-songwriters. Born in Taitung County in the country's southeast, she got her big break when she was spotted singing in a cafe where she also worked as a waitress.

Teng Li Chun, who performed as Teresa Teng, became an Asian superstar in the 1970s and was one of the era's preeminent 'Mandopop' singers. Tragically, she died after an asthma attack when she was only 42 years old.

Playlist

Rainbow
A-Mei
Genre: Pop

Silk Road
A Moving Sound
Genre: World music

Myth
Samingad
Genre: Folk/Pop

Human(ly)
Anie Fann
Genre: Pop

Island's Sunrise
Fire EX
Genre: Indie rock

Your Name Is Taiwan
Kou Chou Ching
Genre: Hip-hop

BIG THING
MJ116
Genre: Hip-hop

Because of You
Mayday
Genre: Pop/Rock

Elder's Drinking Song
Difang Duana
Genre: Folk

Tian Mi Mi
Teresa Teng
Genre: Ballad

Tajikistan

Read List

Shahnameh
by Abolqasem Ferdowsi (1010, trans 1832)
An epic poem from the Persian tradition, with whom Tajiks share a linguistic and cultural heritage. This 'Book of Kings' is a defining pillar of the region's literary culture.

The Sands of Oxus
by Sadriddin Ayni (1998)
This collection of reminiscences from the national poet is a compelling look at the lived history of Tajikistan – and one of the few works of Ayni available in English.

Father of Persian Verse: Rudaki and His Poetry
by Sassan Tabatabai (2010)
This scholarly analysis of Rudaki – the Persian 'Adam of Poets' and progenitor of modern Tajik literature – also offers some of the best available English translations of his work.

The Rubaiyat
by Omar Khayyam (around 1120 CE)
The great Persian mathematician and astronomer delves into poetry with a nature-influenced spirituality that is widely celebrated across the Persian-speaking world – though questions remain as to its authenticity.

Tajiks – Volume 1 & 2
by Bobojon Ghafurov (2011)
This encyclopaedic history of the Tajiks is the defining work for exploring the nation's culture and past – so much so that the author was named a Hero of Tajikistan.

Watch List

The Legend of Rustam
(1971; epic; directed by Bentsion Kimyagarov)
The first of a trilogy (also including *Rustam and Suhrab* and *The Legend of Siyavysche*) based on the Shanameh epic. Classics of Soviet Tajik film.

Odds and Evens
(1993; drama; directed by Bakhtyar Khudojnazarov)
Life is cheap in Dushanbe during the Tajik civil war – a lesson heroine Mira learns quickly when proposed as collateral for her father's gambling debts.

Nisso
(1966; drama; directed by Marat Aripov)
A young girl escapes the hands of a local khan and runs for a Soviet-controlled settlement in this film, set against the gorgeous landscapes of the Tajik Pamir.

Luna Papa
(1999; drama; directed by Bakhtyar Khudojnazarov)
This international collaboration represents the best of post-independence Tajik cinema as a family sets off on a bizarre road-trip to save their pregnant daughter's honour.

Mushkilkusho
(2016; drama; directed by Umedsho Mirzoshirinov)
Shot in the Pamiri language, this love story of a Pamiri girl and Russian boy confronts the prejudice of tradition when she returns home pregnant.

Village life in the Pamir mountains is the backdrop to *Nisso* and other Tajik movies.

Devona Shaw is representative of the Falik musical tradition of Southern Tajikistan and Northern Afghanistan. Folk songs like this are central to both formal community events and informal alike, especially among the region's Pamiri ethnic group.

The Shashmaqam musical tradition, which stretches across the cultures of modern Tajikistan and Uzbekistan, has roots in Sufic mysticism. It represents diverse and multiethnic influences derived from Bukhara's cultural importance in the region.

Playlist

Ay Yorum Biyo
Muboraksho
Genre: Rock

Bad-i-Roshan
Manjigol Khojat
Genre: Folk

Devona Shaw
Davlatmand Kholov
Genre: Folk

Zebo ba Zebo
Daler Nazarov
Genre: Pop

The Imperfection of Our World Inheritance
Tolib Shakhindi
Genre: Classical

Sabza
Kibriyo Rajabova
Genre: Pop

My Soul (Jon-e Manam)
Shashmaqam Ensemble
Genre: Folk

Chashmai Salsabil
Zafar Nozim
Genre: Folk

Boom Boom Boom
Tahmina Niyazova
Genre: Pop

Hay Girdi Yoron
Odina Xoshim
Genre: Folk

Thailand

Read List

Sightseeing
by Rattawut Lapcharoensap (2004)
An award-winning and diverse collection of short stories exploring complex Thai themes: love and prostitution, loss of identity, and Thailand's love-hate relationship with tourism.

Four Reigns
by Kukrit Pramoj (1953, trans 1981)
A dynastic multigenerational epic, tracing the lives of a courtier and her children as Thailand morphs from an absolute monarchy into a modern state against a backdrop of two world wars.

Ramakien
author unknown (13th century)
The Thai version of the Hindu epic, the Ramayana, blends elements from South and Southeast Asia history, faith and legend, giving Hanuman, the monkey god, a central, starring role.

Bangkok Wakes to Rain
by Pitchaya Sudbanthad (2019)
A complex narrative of shared history, loss and alienation in past, present and future Bangkok, as a series of seemingly disconnected stories coalesce around a single apartment building.

The Blind Earthworm in a Labyrinth
by Veeraporn Nitiprapha (2015, trans 2018)
Told as a series of soap opera-like vignettes, this ambitious novel follows a fragile, tangled romance through a world complicated by coups, uprisings and dishonesty.

Watch List

Ong-Bak, Muay Thai Warrior
(2003; martial arts; directed by Prachya Pinkaew)
Everything you'd want from a martial arts extravaganza – stolen Buddha statues, criminal kingpins, henchmen, tuk-tuk chases, even a Muay Thai tournament on Khao San Road.

Uncle Boonmee Who Can Recall His Past Lives
(2010; fantasy drama; directed by Apichatpong Weerasethakul)
A curious exploration of mortality from a Buddhist perspective, as a dying man meets the reincarnated spirits of his lost family and contemplates his past and future lives.

Inhuman Kiss
(2019; horror; directed by Sitisiri Mongkolsiri)
An unsettling tale of a young girl whose head detaches at night and flies off in search of victims to feed on, based on traditional Thai folk legends.

Last Life in the Universe
(2003; drama; directed by Pen-Ek Ratanaruang)
A suicidal Japanese librarian and a bereaved Thai woman stumble into a complex relationship that is interrupted by figures from their overlapping pasts.

Tropical Malady
(2004; drama; directed by Apichatpong Weerasethakul)
A strange fusion of two stories – a romance between two men in a country village, and the tale of a lost soldier, taunted by the spirit of a tiger shaman.

Thailand's unique Muay Thai scene provides the narrative of *Ong-Bak, Muay Thai Warrior.*

Pioneers of the 'songs for life' genre – protest-themed folk – Caravan grew out of Thailand's left-leaning student movement. After the 6 October massacre in 1976, the band took refuge in country villages with the exiled Communist Party of Thailand until an amnesty was declared in 1979.

National heartthrob Thongchai 'Bird' McIntyre is arguably Thailand's most successful pop star. He started out as a bank clerk before being discovered by TV mogul Kai Varayuth and propelled to stardom by a lead role in the 1990 WWII drama *Khu Kam.*

Playlist

Patiroop
Rap Against Dictatorship
Genre: Rap

Ok Huk
Bodyslam
Genre: Rock

Khon Kap Khwai
Caravan
Genre: Phleng phuea chiwit ('songs for life')

Made in Thailand
Carabao
Genre: Rock

Som Sarn
Loso
Genre: Rock

Kee Heung
Silly Fools
Genre: Metal

Sabai Sabai
Bird McIntyre
Genre: Pop

Rak Khun Khao Laeo
Suthep Wongkamhaeng
Genre: Luk krung ('child of the city')

Krung Thep Mahanakhon
Asanee-Wasan
Genre: Rock

My Bloody Valentine
Tata Young
Genre: Pop

Uzbekistan

Read List

Days Gone By
by Abdullah Qodiriy (1925, trans 2019)
Uzbeks' first novel, *'O'tgan kunlar'* by one of the country's revered (then eventually executed) intellectuals, examines the values to be learned from Uzbek traditions in the immediate aftermath of the Bolshevik Revolution.

Hamsa
by Alisher Navoi (1485)
Hamsa was the first work of diwan poetry published in the Chagatai language – the precursor to modern Uzbek – instead of the era's more prevalent Persian. It remains a foundation of Uzbek literary identity.

Baburnama
by Zahiruddin Muhammad Babur (1500s)
A descendant of Tamurlane and Chinggis Khan and founder of the Mughal Dynasty narrates the family's flight from Uzbekistan and establishment in India in an evocative description of life in the early 16th century.

The Devil's Dance
by Hamid Ismailov (2018)
This fictionalised account of Abdullah Qodiriy's arrest and final days in prison celebrates Qodiriy's creativity and final lost novel while condemning the repressive regime that cut his career tragically short.

Night and Day
by Cholpon (1934, trans 2019)
In the first of a two-part series left unfinished by Stalinist purges, one of Uzbekistan's most-celebrated authors explore the realities of life in Soviet Central Asia.

Watch List

The Orator
(1998; drama; directed by Yusup Razikov)
In *'Voiz'*, religious traditions and the march of Soviet revolution combine to turn protagonist Iskender's life upside down – losing his own identity along the way.

Delighted By You
(1958; comedy; directed by Yo'ldosh A'zamov)
The first Uzbek comedy, *'Maftuningman'* is equally delightful for the lighthearted theme and honest looks it offers into the rural Uzbek culture of the time.

Without Fear
(1972; drama; directed by Ali Khamraev)
As Soviet power envelops remote Uzbek villages, opposing cultures collide at the intersection of propriety and progress – with deadly results for the key characters.

Parizod
(2012; drama; directed by Ayub Shahobiddinov)
The appearance of a mysterious woman in a rural Uzbek village brings out the worst aspects of human emotion and vice – despite her angelic aura.

Hot Bread
(2018; drama; directed by Umid Khamdamov)
Young protagonist Zulfiya longs for the big city life and the warmth of her far-away mother, while life in the village is oppressive and unchanging.

Hot Bread was the Uzbek entry for Best International Film at the 92nd Academy Awards.

Playlist

Kechalar
Shahzoda
Genre: Pop

Samarqand Ushoq
Sadriddin Gulov
Genre: Folk

Uchkuduk
Yalla
Genre: Pop/Rock

One of the first Uzbek super-groups, Yalla was a massive hit all across the Soviet Union and allied countries at their peak in the 1980s. The band remains popular with Uzbeks around the world.

Bilmaysan
Lola Yuldasheva
Genre: Pop

Särbinaz
Gülnara Allambergenova & Injigul Saburova
Genre: Folk

Yurak Mahzun
Sogdiana
Genre: Pop

STIHIA Experiment
KEBATO
Genre: Techno

Founder of the groundbreaking Stihia electronic festival deep in the Uzbek deserts, DJ KEBATO is singlehandedly changing the Uzbek music scene, causing a ripple effect across Central Asia and the former USSR. His real name is Otabek Suleimanov – he's also a hot-shot lawyer.

Arabian Tango
Botyr Zokirov
Genre: Tango

Baxtli bo'laman
Rayhon
Genre: Pop

Galdir
Nodira Pirmatova
Genre: Folk

Vietnam

Read List

Dumb Luck
by Vu Trong Phung (1936, trans 2002)
A satire of the modernisation of 1930s Hanoi and the rise of a vagabond to the French upper crust. Its parody of colonial capitalism had the book banned in Vietnam until 1986.

On Earth We're Briefly Gorgeous
by Ocean Vuong (2019)
Written as a letter, Vuong's lyricism transports you across time and generations, from Connecticut as an immigrant to his mother's Vietnam. Bursting with poetry, the impressive novel explores how to survive with grace.

Such a Lovely Little War: Saigon 1961-63
by Marcelino Truong (2016)
The son of a Vietnamese diplomat guides us through Saigon (and the USA) across love and war, culture-clash and his French mother's bipolar disorder in a smart graphic novel.

The Stars, The Earth, The River
by Le Minh Khuê (1997)
Khuê peers unflinchingly into life in Vietnam before, during and after the 'American war', through the eyes of a female Vietnamese war veteran. Moments of black humour brighten daily life and war atrocities.

Ticket to Childhood: A Novel
by Nguyen Nhat Anh (2008, trans 2014)
You can almost hear the southern Vietnamese lilts in this voice of a childhood in Saigon. This novel about recapturing youthful creativity was a hit with local readers.

Watch List

The Scent of Green Papaya
(1993; drama; directed by Tran Anh Hùng)
The natural world of Vietnam is palpable through birdsong and colourful produce in this exquisite film about an orphan girl falling dangerously in love in pre-war Saigon.

Nuoc (2030)
(2014; sci-fi; directed by Nguyen Vo Nghiem Minh)
Water looks equally dreamy and chilling in a future Vietnam where half of the country has been swallowed up by the sea through climate change. At its heart is a headstrong Vietnamese woman battling the loss of love and land as she escapes to a floating farm.

Ao lua Ha Dong
(2006; drama; directed by Luu Huynh)
In '*The White Silk Dress*', Vietnam's traditional dress becomes a power symbol of the tenacity, yet enduring elegance, of three Vietnamese women through the war.

Thang Nam Ruc Ro
(2018; rom-com; directed by Nguyen Quang Dung)
Vietnam bursts with pastel joy in '*Go-Go Sisters*', a film about getting the old highschool girl-gang back together. Characters break into song, capturing contemporary Vietnam's youthful verve.

When the Tenth Month Comes
(1984; drama; directed by Dang Nhat Minh)
Once a year, the living and deceased mingle at a 'ghost market'. These touches of magic realism paint a woman's grief for her husband, killed in the Vietnam War.

The melodic voice of Trung Kien Trinh navigates urban melancholy and yearning. He is a vibrant part of the indie-pop scene coming out of Hanoi and Ho Chi Minh City, using music to narrate Vietnamese life.

Vietnamese hip-hop boasts less about wealth and instead whispers, spits and croons the emotions and lifestyles of ordinary Vietnamese people. Suboi raps about Ho Chi Minh traffic over erratic, urgent beats.

Many a movie and book offer accounts of the American War – and try to make sense of it.

Playlist

Tinh Thoi Xot Xa
Lam Truong
Genre: Easy listening

The Ki 21 Buon
Trung Kien Trinh
Genre: Indie pop

Co Em Cho
MIN
Genre: Pop

Mot Cuoc Song Khac
Empty Spaces
Genre: Prog rock

Dinh Nui Tuyet Cua Nuoi Tiec
Datmaniac
Genre: Rap

Tu Phu
Hoang Thuy Linh
Genre: New Age

Chung Ta Khong Thuoc Ve Nhau
Son Tung M-TP
Genre: R&B/Pop

Nghin Trung Xa Cach
Thai Thanh
Genre: Folk/Opera

N-SAO?
Suboi
Genre: Hip-hop

Khoang Khong Bo Lai
KOP band
Genre: Alt rock

AFRI
MIDDL

CA &
E EAST

Angola

Read List

The World of Mestre Tamoda
by Uanhenga Xitu (1988)
Xitu's only English translation was written while he was in prison in Portugal and contains three stories centred on the character of Tamoda, a loquacious bush lawyer exploring the urban-rural divide in colonial Angola.

Yes, Comrade
by Manuel Rui (1977)
Set in the 1960s and '70s revolutionary struggles, this collection of short stories is the only English translation by one of Angola's finest post-independence writers.

Luuanda
by José Luandino Vieira (1963)
Ground-breaking short-story collection that critiqued Angola's Portuguese administrators and had the dual 'honour' of winning two literary awards while being banned by Portugal's repressive Salazar government.

Transparent City
by Ondjaki (2012)
Ondjaki, aka Ndalu de Almeida, one of Angola's most widely translated authors, won the prestigious José Saramago Prize for this depiction of post-civil war Luanda populated with a rich cast of characters.

Sacred Hope
by Agostinho Neto (1974)
Angola's first president was something of a Renaissance man, studying medicine in Portugal but also becoming an acclaimed poet. This collection contains poems that have been turned into national anthems.

Watch List

Na Cidade Vazia
(2004; drama; directed by Maria João Ganga)
A landmark of Angolan cinema and the first film to be directed by a woman, '*Hollow City*' highlights the plight of an orphan arriving in a Luanda at the end of the civil war.

O Herói
(2004; drama; directed by Zézé Gamboa)
Released in the wake of the civil war, this Angolan–Portuguese–French feature examines the conflict's after-effects through the lives of four characters: a war veteran, a 10-year-old boy, a teacher and a prostitute.

Sambizanga
(1972; drama; directed by Sarah Maldoror)
A politically charged film about Angola's independence struggle that was made in Congo while the war was still raging. French director Maldoror's husband was a key player in the anti-colonial movement.

Death Metal Angola
(2012; documentary; Jeremy Xido)
Wonderfully esoteric film about an underground youth movement trying to organize a death metal concert in the bombed out Angolan city of Huambo. Beautifully shot, it quickly garnered critical acclaim and a cult following.

Ar Condicionado
(2020; drama; directed by Fradique)
Artistically photographed film set in a scarred tenement building in downtown Luanda. It explores the 'invisible' people of the city, including security guard Matacedo and his mission to retrieve a mysterious air-conditioner.

Playlist

Angola, País Novo
Matias Damásio
Genre: Kizomba

Mona Ki Ngi Xica
Bonga
Genre: Semba

Njila ia Dikanga
Paulo Flores ft Yuri da Cunha
Genre: Semba

Born in Luanda and partly raised in Portugal, Flores is a master of contemporary *semba*, a form of Angolan music and dance from the 1940s still popular for its syncopated rhythms and mash-up of African and European influences.

Meninos do Huambo
Ruy Mingas
Genre: Kizomba

War Heads
Neblina
Genre: Metal

Kuduro Luvin
Coréon Dú ft Phil Asher
Genre: Kuduro

Kamikaze Angolano
Ikonoklasta
Genre: Rap

Ikonoklasta, aka Luaty Beirão, is the son of a former big-shot in the Angolan government. Known as much for his political activism as his music, he has spent time in jail after protesting about corruption, inequality and press freedom in Angola.

Ndatekateka
Justino Handanga
Genre: Semba

Makalakatu
Paulo Flores
Genre: Semba

Wanga
Angela Ferrão
Genre: Semba

The 2020 movie *Ar Condicionado* was set and filmed in the Angolan capital Luanda.

Africa & Middle East

Benin

Read List

Why Goats Smell Bad and Other Stories from Benin
by Raouf Mama (1997)
Benin has a great folklore tradition, reflected by this illustrated collection of 19 Fon stories. Retold in English by Mama, they cover orphans and twins with magical associations, spirits, animals, royalty and farmers.

Butterfly Fish
by Irenosen Okojie (2014)
Longlisted for the Man Booker Prize, Okojie's magical-realist debut weaves together contemporary London and a vividly imagined 18th-century Benin, connected by a brass-cast Beninese warrior's head.

Spirit Rising: My Life, My Music
by Angélique Kidjo (2014)
Benin's most famous export tells the story of her perilous escape from communist Benin to France, and her subsequent rise to winning four Grammy awards and topping Billboard's World Album chart.

The Viceroy of Ouidah
by Bruce Chatwin (1980)
The late English travel writer's fictionalised sketch of the notorious Brazilian slave trader Francisco da Silva, who traded with the kings of Dahomey (precolonial Benin).

Show Me the Magic: Travels Round Benin by Taxi
by Annie Caulfield (2002)
Travelling in Isidore's taxi from colourful southern Benin to the Islamic, sub-Saharan north, the late British travel writer and broadcaster discovers a land characterised by corruption, Catholicism and *gris-gris* (voodoo).

Watch List

In Search of Voodoo: Roots to Heaven
(2018; documentary; directed by Djimon Hounsou)
Cotonou-born Hollywood actor Hounsou returns to his homeland to study the animistic roots of voodoo, which is still practised in Benin. Voodoo travelled on the slave ships to the Americas, where it took a darker form.

Cobra Verde
(1987; drama; directed by Werner Herzog)
A German-directed dramatisation of Chatwin's *The Viceroy of Ouidah*, shot in West Africa and Latin America. Klaus Klinksi plays the bandit who leads an army of female warriors against Dahomey's King Bossa.

Cocoa, Slaves, & Goo
(2005; documentary; directed by John Maguire)
In this episode of *Geldof in Africa*, in which the musician returns to the continent two decades after Live Aid, he travels the old slave road from Ghana to Benin.

Africa Paradis
(2006; drama; directed by Sylvestre Amoussou)
A satirical look at the issue of African emigration, in which a French couple try to immigrate to a thriving African country. Actor and director Amoussou is Beninese.

Bight of the Twin
(2016; documentary; directed by Hazel Hill McCarthy III)
As much a musical soundscape as a documentary, built around the experimental work of Genesis P-Orridge and his industrial band Throbbing Gristle, this film explores Beninese voodoo.

Afropop star Angélique Kidjo has won several Grammy awards for global music.

Going since the 1960s, the Orchestre Poly-Rythmo de Cotonou are funk supremos from Benin's de facto capital, often referred to as Tout Puissant ('all mighty'). They incorporate the rhythms of voodoo ceremonial music.

Benin-born guitarist Lionel Loueke has collaborated with the likes of singer Angélique Kidjo, Australia's The Vampires and US pianist Herbie Hancock. The latter was Loueke's mentor, who praised him as 'a musical painter'.

Playlist

Helou
Sagbohan Danialou
Genre: Percussive jazz/Traditional

Wombo Lombo
Angélique Kidjo
Genre: Afropop

Yiri Yiri Boum
Gnonnas Pedro
Genre: Highlife/Jùjú

Mi Si Ba To
Orchestre Poly-Rythmo de Cotonou
Genre: Soukous/Funk

Yoruba
Gangbé Brass Band
Genre: Afrobeat/Voodoo

Gbé
Lionel Loueke
Genre: Jazz

You Are My Dream
Nel Oliver
Genre: Akpala/Funk

Pour Son Amour
Zeynab Habib
Genre: Afropop

Huankpe Dje Mi
Yelouassi Adolphe
Genre: Afro-Cuban

Qu'est-ce qu'on n'a jamais vu
Blaaz
Genre: Hip-hop

Cape Verde

Read List

The Madwoman of Serrano
by Dina Salústio (2012, trans 2020)
This inventive magic realism novel – the first by a female author to be published in Cape Verde and also the first to be translated into English– was shortlisted for the Oxford-Weidenfeld Translation Prize 2020.

The Last Will & Testament of Senhor da Silva Araújo
by Germano Almeida (1989, trans 2004)
Taking as its starting point the unusual reading of a businessman's will, this lively novel throws a witty spotlight on life on São Vicente and the archipelago.

Cape Verdean Blues
by Shauna Barbosa (2018)
You'd be hard-pressed to find a better collection of words that express the idea of roots and identity than this American-Cape Verdean poet's beautiful, unforgettable debut collection.

Chiquinho
by Baltazar Lopes (1947, trans 2019)
Considered by many to be the greatest Cape Verdean novel, this richly engaging coming-of-age tale is also a compelling exploration of the archipelago's unique identity.

Other American Dreams
by Sérgio Monteiro (2015)
Cape Verde's role in slavery and its history of migration lie at the heart of this intense thriller about a boat full of dead migrants washing up on the shores of a sleepy island nation.

Watch List

Djon Africa
(2018; drama; directed by João Miller Guerra, Santa Martha & Filipa Reis)
This documentary film delivers a luscious slice of island life with its warm-hearted tale about a young idler trying to track down his father.

The Island of Contenda
(1996; drama; directed by Leão Lopes)
Henrique Teixeira de Sousa's much loved novel about progress and change on Cape Verde is vividly brought to life via the landscapes of Fogo island and the songs of Cesária Évora.

Down to Earth
(1995; drama; directed by Pedro Costa)
There's not much of a narrative to this tale about a Portuguese nurse accompanying a comatose patient back to Cape Verde, but the landscape of melancholic volcanoes delivers an unforgettable experience.

Napomuceno's Will
(1997; drama; directed by Francisco Manso)
Adapted from *The Last Will & Testament of Senhor da Silva Araújo* by Germano Almeida (see read list), this engaging movie about the aftermath of a will reading has shades of Almodovar in its witty narrative.

Fintar o Destino
(1998; drama; directed by Fernando Vendrell)
The dilemmas and decisions at the heart of migration are explored beautifully in this film, *'Dribbling Fate'*, about a washed-up goalkeeper who's given a second chance.

The unique landscape of the Cape Verde islands informs many of its books and films.

Similar to Portuguese *fado* but with a local African beat, *mornas* are slow, melancholic folk songs usually delivered by throaty female vocalists. They're the sound of Cape Verde, and their most famous exponent was Cesária Évora.

Funaná is everywhere on Cape Verde; in bars, cafes and clubs people gather to listen to the raucous, high-energy melodies of a genre unique to the island, played on the accordion-like *ferrinho*. Once banned for sounding too 'African', its reintroduction in 1975 was a welcome symbol of the island nation's new independent status.

Playlist

Nha berçu
Ildo Lobo
Genre: Morna

Sema Lopi
Bulimundo
Genre: Funaná

Song for my Father
Horace Silver Quintet
Genre: Jazz

6 on na Tarrafal
Carmen Souza
Genre: Traditional

Sodade
Cesária Évora
Genre: Morna

Bidibido
Elida Almeida
Genre: Funaná

Tabanca
Orlando Pantera
Genre: Batuque

Il Paraiso
Suzanna Lubrano
Genre: Zouk

Dor de Mundo
Camilo Domingos
Genre: International

Tunuka
Mayra Andrade
Genre: World music

Democratic Republic of Congo

Read List

Tram 83
by Fiston Mwanza Mujila (2014, trans 2015)
In Mujila's award-winning debut, a budding writer meets his racketeer friend in a notorious nightclub, plunging the reader into a darkly comic evocation of the modern African gold rush.

Before the Birth of the Moon
by VY Mudimbe (1976, trans 1989)
Set shortly after the DRC's independence from Belgium, this story throws 'The Minister' and prostitute Ya together in 1960s Kinshasa's nightclubs and bars.

Mama Koko and the Hundred Gunmen
by Lisa J Shannon (2014)
The American human rights activist and author of A Thousand Sisters travels with Congolese expatriate Francisca Thelin to her homeland in the shadow of Joseph Kony's brutal Lord's Resistance Army.

The Poisonwood Bible
by Barbara Kingsolver (1998)
Kingsolver's Pulitzer-nominated novel tells the story of an American missionary family in the Belgian Congo, covering the Congo Crisis that followed the country's independence in 1960.

Johnny Mad Dog
by Emmanuel Dongala (2002, trans 2006)
The Congolese writer and chemist prefaced books and films such as *Beasts of No Nation* with his story of child soldiers in a war-torn West African country. It was later adapted as a film set in Liberia.

Watch List

Downstream to Kinshasa
(2020; documentary; directed by Dieudo Hamadi)
The DRC's first inclusion in the Cannes Film Festival follows survivors of the Six-Day War of 2000, as they travel to the capital to demand government compensation for the losses they incurred.

Identity Pieces
(1998; comedy; directed by Mwezé Ngangura)
This international film festival favourite follows the misadventures of a Bakongo king who travels to Brussels to find his daughter, encountering European prejudices and the African diaspora.

This is Congo
(2017; documentary; directed by Daniel McCabe)
This beautifully shot documentary by an American filmmaker covers the DRC's ongoing Kivu conflict and humanitarian crisis, taking in past and present, equatorial landscapes and chaotic streets.

Félicité
(2017; drama; directed by Alain Gomis)
Featuring the Kasai Allstars band, this freeform film follows singer Félicité as she searches for her teenage son in nocturnal Kinshasa.

Viva Riva!
(2010; thriller; direccted by Djo Tunda Wa Munga)
This crime thriller sees fuel smuggler Riva pursued through the streets of Kinshasa by various sinister characters, while he complicates matters further by falling for a gangster's girlfriend.

Author Emmanuel Dongala's family left their home in Brazzaville when war swept the DRC.

Melodic *soukous* developed out of Congolese rumba, thanks to TPOK Jazz and Zaïko Langa Langa. Formed in 1969, the first word in this long-running band's name – Zaïko – is a portmanteau referring to Zaire, the DRC's name from 1971 to 1997.

This edgier take on *soukous* is one of the DRC's latest pan-African musical exports. Werrason is one of *ndombolo*'s biggest names, along with the R&B-tinged singer-songwriter Fally Ipupa.

Playlist

Masanga
Jean Bosco Mwenda
Genre: Fingerstyle guitar

Mabeley a mama
Wendo Kolosoy
Genre: Soukous

Attention na SIDA
Franco Luambo & TPOK Jazz
Genre: Congolese rumba

Linya
Zaïko Langa Langa
Genre: Soukous

Indépendance Cha Cha
Le Grand Kallé & L'African Jazz
Genre: Soukous

Block Cadenas
Werrason
Genre: Ndombolo

Mobali Na Ngai Wana
Tabu Ley Rochereau, M'bilia Bel & Afrisa International
Genre: Congolese rumba

Monie
Kanda Bongo Man
Genre: Kwassa kwassa

Paradiso
Konono No1
Genre: Bazombo trance

Hallo
DRC Music
Genre: Trip-hop

Egypt

Read List

The Cairo Trilogy
by Naguib Mahfouz (1956, trans 1990)
This epic of modern-Egyptian literature by Nobel laureate Mahfouz traces the fate of one Cairo family, alongside Egypt's sweeping societal changes, in the era of British rule.

The Open Door
by Latifa al-Zayyat (1960, trans 2000)
Set in the last decades of British colonial control, Latifa al-Zayyat's Egyptian feminist classic follows Layla as she struggles to find her identity amid Egypt's fight for independence.

The Yacoubian Building
by Alaa Al Aswany (2002, trans 2004)
This local and international bestseller portrays the lives and loves of the occupants of one downtown Cairo building as they contend with a society suffused in political corruption and religious hypocrisy.

Taxi
by Khaled Alkhamissi (2006, trans 2011)
Filled with both wit and anger, Khaled Alkhamissi's novel of fictional monologues with Cairo's taxi drivers shines a light on the hard scrabble of day-to-day life for average Egyptians.

Brooklyn Heights
by Miral al-Tahawy (2010, trans 2011)
Miral al-Tahawy weaves a tale of the Egyptian immigrant experience in this novel, which moves between New York City and a village in the Nile Delta.

Watch List

The Will
(1939; drama; directed by Kamal Selim)
A young couple battle to hold on to their relationship amid the economic crisis of 1930s Egypt in this black-and-white classic of Egyptian cinema.

The Night of Counting the Years
(1969; drama; directed by Shadi Abdel Salam)
This film is based on the true story of Luxor's Royal Cache tomb robberies and the hunt to find out who was selling its antiquities.

Terrorism & Kebab
(1992; comedy; directed by Sherif Arafa)
Set in Cairo's reviled Mogamma building (the city's bureaucracy hub), this beloved black-comedy follows Ahmed as he takes the Mogamma hostage after failing to renew his paperwork.

Cairo 678
(2010; drama; directed by Mohamed Diab)
This drama tracks the individual stories of three Egyptian females as they endure day-to-day sexual harassment and endeavour to seek justice against the perpetrators.

Excuse My French
(2014; comedy; directed by Amr Salama)
Religious and class differences in Egyptian society are at the fore of this comedy about Hany, a 12-year-old Christian who is mistaken for Muslim at his new school.

Cairo's changing culture is described in Naguib Mahfouz's *The Cairo Trilogy*.

Legendary singer Umm Khalthum – described as 'Egypt's fourth pyramid' – had a singing and acting career spanning 50 years and remains iconic in the Arab world. Some of her songs, performed live, could last up to 90 minutes.

Mahraganat is a style of street music that fuses elements of hip-hop and electronica with Egyptian *shaabi* (the rhythmic music of the working-class, that itself derived from traditional Egyptian folk).

Playlist

Zarhat el Louxor
Les Musiciens du Nil
Genre: Sa'idi folk

Habibi
Ali Hassan Kuban
Genre: Modern Nubian

Nour El Ein
Amr Diab
Genre: Arabic pop

Zahma Ya Dunya Zahma
Ahmed Adaweyeh
Genre: Shaabi

Enta Omry
Umm Khulthum
Genre: Classical Arabic

Basha E3temed
Abyusif
Genre: Hip-hop

Hitta Minni
Hysa, Halabessa & Sweasy
Genre: Mahraganat

Ya El Medan
Cairokee
Genre: Rock

Baladi Dance
Abu Zariya Band
Genre: Baladi folk

Ala Bali
Sherine
Genre: Egyptian pop

Ethiopia

Read List

Love to the Grave
by Haddis Alemayehu (1968)
Former Foreign Minister Alemayehu's tragic story of love and class has been described as the country's *Romeo and Juliet*, and is one of the great works of modern Ethiopian literature.

Notes From the Hyena's Belly
by Nega Mezlekia (2000)
Mezlekia's moving memoir starts with a happy childhood in the multicultural eastern city of Jijiga, before darkness comes in the form of revolution, famine and war with neighbouring Somalia.

Cutting for Stone
by Abraham Verghese (2009)
This absorbing bestseller follows the fortunes of twin brothers who grow up to be surgeons, in a book that digs deep into both medical procedure and Ethiopia's troubled late 20th century.

Children of the Revolution
by Dinaw Mengestu (2007)
Mengestu – himself an Ethiopian migrant to the US – offers the powerful, tightly told account of a man who flees the Derg regime and ends up running a convenience store in Washington, DC.

Beneath the Lion's Gaze
by Maaza Mengiste (2010)
Tells a poignant story of a fictional family during the last days of Haile Selassie, by now confined to a room with his pet lions for company, and the rise of the Derg junta.

Watch List

Town of Runners
(2012; documentary; directed by Jerry Rothwell)
The remote highland town of Bekoji has produced some of the world's best distance runners. This intriguing documentary follows two teenagers as they battle to join the elite.

Difret
(2014; drama; directed by Zeresenay Berhane Mehari)
This earnest drama, based on a true story, explores a lawyer's battle to defend a girl charged with shooting a man who had tried to kidnap her for marriage.

Lamb
(2015; drama; directed by Yared Zeleke)
This story of a boy who moves in with his relatives when his mother dies was the first Ethiopian film to make the Official Selection at Cannes. It gives a lovely flavour of everyday life, in all its hardship and joy.

Price of Love
(2015; drama; directed by Hermon Hailay)
A cab driver helps a girl home from a club, earning the enmity of her pimp, who steals his taxi in revenge, in this romantic drama set in Addis Ababa.

Hairat
(2017; documentary; directed by Jessica Beshir)
Beshir grew up in Harar, and this ghostly short follows Yussuf Mume Saleh, who feeds hyenas at night near the city walls.

Astatke is known as the father of Ethio-jazz for his combination of Ethiopian music (which uses a five-note scale and often asymmetrical rhythms) with jazz and Latin. Astatke studied in Europe and the USA before returning to Ethiopia in the 1960s, and his shifting, hypnotic music remains popular today.

The hyena feeders of Harar are featured in the short documentary *Hairat*.

Instruments such as lutes, lyres, gongs and drums are used in traditional Ethiopian music. Islamic and Christian songs are also common. The Dorze of Southern Ethiopia are famous for their cotton weaving, beehive-like huts and polyphonic singing, which can feature five overlapping vocal parts.

Playlist

Ethiopia
Teddy Afro
Genre: Reggae

Gera Geru
Gigi
Genre: Ethiopian music

Yezebarekalu
Eyob Mekonnen
Genre: Reggae

Fiker Beamargna
Abdu Kiar
Genre: Reggae

Yègellé Tezeta
Mulatu Astatke
Genre: Ethio-jazz

Polyphony Dorze
Dorze Chorus
Genre: Traditional

Tew Limed Gelaye
Mahmoud Ahmed
Genre: Tizita

Homesickness, pt 2
Tsegué-Maryam Guèbrou
Genre: Blues

Shellèla
Getatchew Mekurya
Genre: Ethio-jazz

Fikir
Aster Aweke
Genre: Ethiopian music

The Gambia

Read List

Reading the Ceiling
by Dayo Forster (2007)
On the cusp of womanhood, Ayodele must choose between three different suitors, each leading to a dramatically different future. The three-part story explores how events effect who we become.

The Sun Will Soon Shine
by Sally Singhateh (2004)
Though a gifted student with a bright future, Nyima must bow to traditional pressure and marry a cruel man from her village – a harrowing ordeal faced by countless women around the globe.

Chaff on the Wind
by Ebou Dibba (1986)
Set in the 1930s, two ambitious young Gambians move from village to city in pursuit of opportunity and adventure. Vivid settings (markets, villages, docks) star in this Dickensian page-turner.

Dream Kingdom
by Tijan M Sallah (2007)
One of Africa's most talented poets delves into a world of folklore, African politics, global injustice and more in this collection of mystical works penned over 25 years.

Roots: The Saga of an American Family
by Alex Haley (1976)
This Pulitzer Prize-winning masterpiece shows the horrors of slavery over many generations. It begins with a violent raid on the Gambian village of Jufureh, where author Alex Haley traced his roots.

Watch List

Gambia: Take Me to Learn My Roots
(2019; documentary; directed by Bacary Bax)
Two young men from Hull, England, return to their childhood roots in West Africa. Vibrant streetscapes and a first-rate soundtrack provide the backdrop.

Kings of the Gambia
(2010; documentary; directed by David Vogel)
The mesmerising *kora* playing and lush vocals of Lamin Jobarteh take centre stage as an African-Swiss band travel the River Gambia, giving village concerts along the way.

The Mirror Boy
(2011; drama; directed by Obi Emelonye)
This redemptive tale of self-discovery features an African-English boy travelling to his mother's Gambian homeland on a journey filled with mysterious encounters.

Jaha's Promise
(2017; documentary; directed by Patrick Farrelly and Kate O'Callaghan)
The portrayal of a courageous Gambian woman seeking to end the barbaric practice of female genital mutilation. Jaha Dukureh's campaign against ignorance, violence and misogyny shows that one person can bring great change.

Beyond, An African Surf Documentary
(2017; documentary; directed by Mario Hainzl)
Inspiring depiction of homegrown surfers in the Gambia and other West African countries. Gorgeous photography and fascinating insight into the culture clash between individualism and traditional values.

Author Alex Haley, here in 1977, partly wrote *Roots* during the night on cargo ships.

Jobarteh hails from a long line of Griots – traditional West African musicians, poets and storytellers. She is also one of the few women virtuosos of the *kora* – a 21-stringed harplike instrument typically passed from father to son.

Beloved in both The Gambia and in Senegal, Moussa Ngom was one of the pioneers of *mbalax*, a style of West African dance music that incorporates jazz, Latin, funk and Congolese rhythms.

Playlist

Jooka Tamala
Foday Musa Suso
Genre: Folk

Serekunda
Jali Madi Kanuteh
Genre: Folk

Yamaro
OBoy & Gambia Child
Genre: Folk

Turn by Turn
Jizzle
Genre: Afropop

Gambia
Sona Jobarteh
Genre: Folk

Fahass
Bai Babu
Genre: Mbalax

Gal Gui
Moussa Ngom
Genre: Mbalax

Brikama
Nana Mboob
Genre: Folk

Hakatu Mas
Jaliba Kuyateh
Genre: Folk

Nna Jarabi
Nobles Gambia
Genre: Afropop

Ghana

Read List

Harmattan Rain
by Ayesha Harruna Attah (2008)
Attah's debut follows three generations of Ghanaian women from the pre-independence Gold Coast to New York: a female view of 20th-century Ghanaian history.

The Beautyful Ones Are Not Yet Born
by Ayi Kwei Armah (1968)
A nameless man works as a clerk trying to feed his family in this debut by Chinua Achebe's near-contemporary, depicting life for the Ghanaian everyman during President Kwame Nkrumah's repressive regime.

My First Coup d'Etat
by John Dramani Mahama (2012)
The former Ghanaian president's coming-of-age memoir covers the 'lost decades' that followed Africa's hopeful independence from colonialism. Mahama's politician father was imprisoned in the coup that ousted Nkrumah.

Our Sister Killjoy
by Ama Ata Aidoo (1977)
Also titled *Reflections from a Black-eyed Squint*, the former Minister of Education's debut novel examines the African diaspora through a young Ghanaian woman's visit to Germany and England.

Homegoing
by Yaa Gyasi (2016)
The Stanford-educated Ghanaian American novelist's bestselling debut was one of Oprah Winfrey's top 10 books in 2016. It traces the slave trade from the 18th-century Gold Coast through to Jazz Age Harlem.

Watch List

Keteke
(2017; comedy; directed by Peter Sedufia)
Available on Netflix, this '80s period drama follows a couple who are determined to give birth in their hometown, prompting an adventure through rural Ghana after they miss the weekly train.

Azali
(2018; drama; directed by Kwabena Gyansah)
Also on Netflix, this harrowing drama chronicles 14-year-old Amina's escape from an arranged marriage into poverty and sex work in Accra. It was Ghana's first submission for the Best International Feature Film Oscar.

The Witches of Gambaga
(2010; documentary; directed by Yaba Badoe)
Made over five years, this international film festival favourite tells the disturbing story of a community of women condemned as witches and exiled to a village in northern Ghana.

The Burial of Kojo
(2018; drama; directed by Blitz Bazawule)
Magical-realist drama by the Ghanaian–American rapper (stage name Blitz the Ambassador) and co-director of Beyoncé's Grammy-nominated *Black Is King*. It was Ghana's first Golden Globes entry and Netflix premiere.

Love Brewed in the African Pot
(1981; romantic drama; directed by Kwah Ansah)
The first privately financed Ghanaian feature film is considered a classic for its *Romeo and Juliet*-like story of love across the social classes during the colonial era.

Born in Ghana, raised in Alabama, Yaa Gyasi's latest book is *Transcendent Kingdom*.

Hiplife is a mix of hip-hop and highlife, Ghana's melodic fusion of jazzy horns, calypso guitar and traditional Akan music. Now a member of VVIP, Rockstone is known as the godfather of hiplife, rapping in Akan, Twi and English.

Ebo Taylor is a veteran of Ghana's highlife and the Afrobeat genre popularised by Nigeria's Fela Kuti. The octogenarian has worked with hip-hop superstars, including Usher, who sampled *Heaven* on his track *She Don't Know* ft Ludacris.

Playlist

Bukom Mashie
Oscar Sulley and the Uhuru Dance Band
Genre: Afrobeat/Funk

Yaa Amponsah
Ogyatanaa Show Band with Pat Thomas
Genre: Highlife

Kwame Nkrumah
E T Mensah & the Tempos
Genre: Highlife

Happy Day
Reggie Rockstone
Genre: Hiplife

Adonai
Sarkodie
Genre: Hip-hop

Heaven
Ebo Taylor
Genre: Highlife

No Dulling
Keche ft Kuami Eugene
Genre: R&B

Konkontiba
Obour ft Batman
Genre: Hiplife

Hiplife
Obrafour ft Sarkodie
Genre: Twi rap

Home Sweet Home
Afro Moses
Genre: Reggae

Iran

Read List

Shahnameh
by Abolqasem Ferdowsi (1010; trans 1832)
'*The Book of Kings*' contains 60,000 verses and tells the mythical – and actual – history of Iran. It's considered a time capsule of Iranian national identity.

The Blind Owl
by Sadegh Hedayat (1936, trans 1958)
Written during the oppressive latter years of Reza Shah's rule (1925–1941), this masterwork by the father of Persian magical realism is a feverish tale of a painter's descent into the nightmare of his own delusions.

Persepolis
by Marjane Satrapi (2000)
As the great-granddaughter of Iran's last emperor, Satrapi bears witness to a childhood uniquely entwined with the history of her country in this semi-autobiographical graphic novel.

The Story of Leyla and Majnun
by Nizami Ganjavi (around 1192, trans 1966)
The most famous telling of a celebrated love story, this epic by the 12th-century Persian poet is thought to have inspired everyone from Shakespeare (*Romeo and Juliet*) to Eric Clapton (*Layla*).

Daiey Jan Napoleon
by Iraj Pezeshkzad (1973, trans 1996)
Set during the Allied occupation of Iran in WWII, '*My Uncle Napoleon*' follows a 13-year-old as he observes his dysfunctional family and neighbours, in particular the eccentric family patriarch, uncle Napoleon.

Watch List

A Separation
(2011; drama; directed by Asghar Farhadi)
In this Academy Award-winning film, a married couple are torn between leaving Iran to improve the life of their child and staying in the country to care for a parent with Alzheimer's.

This Is Not a Film
(2011; documentary; directed by Mojtaba Mirtahmasb & Jafar Panahi)
Jailed Iranian director Jafar Panahi's zero-budget film about a day in his life, shot entirely in his flat, was smuggled out of Iran and screened at Cannes.

Offside
(2006; comedy-drama; directed by Jafar Panahi)
Highlighting the injustice of Iran's ban on women entering sports stadiums, this film sees a group of Iranian girls arrested for trying to enter a stadium dressed as boys in order to watch a big football match.

The Cow
(1969; drama; directed by Daryush Mehrjui)
Widely considered to be the first film of Iranian New Wave cinema, the story of the relationship between villager Masht Hassan and his beloved cow is known for its psychological and social criticisms.

Bashu, The Little Stranger
(1989; drama; directed by Bahram Beizai)
A young boy who loses his house and family in the Iran–Iraq war forms an unlikely bond with a mother, Naii, in a different part of Iran, despite their cultural differences.

Female Iranian football fans strive to be permitted to watch the sport in stadiums.

Forced to quit performing following the Islamic Revolution of 1979, the artist known as Googoosh didn't appear on stage again until she left Iran in 2000. In that year, she performed nearly 30 concerts in Europe and North America.

In late 2009, Sirvan Khosravi's *Saat-e 9* – '9 O'Clock' – became the first domestic post-Islamic Revolution Iranian song to achieve high-rotation airplay on European radio.

Playlist

Nedaye Eshgh
Mohammad Reza Shajarian
Genre: Traditional folk

Banoo
Sima Bina
Genre: Traditional folk

Zolf
Mohsen Namjoo
Genre: Blues/Jazz/Folk

Mahtab
Vigen Derderian
Genre: Pop

Mahkloogh
Googoosh
Genre: Pop

Saat-e 9
Sirvan Khosravi
Genre: Pop

Soghati
Hayedeh
Genre: Pop

Koodakaneh
Farhad
Genre: Folk

Afsoongar
Arian Band
Genre: Pop

Beni Beni
Niyaz
Genre: Folk

Israel & the Palestinian Territories

Read List

Waking Lions
by Ayelet Gundar-Goshen (2014, trans 2016)
A hit-and-run accident in the Negev desert involving an Israeli doctor and an Eritrean refugee sets off this novel about immigration, privilege and guilt, written by novelist and psychologist Gundar-Goshen.

Suddenly, a Knock on the Door
by Etgar Keret (2010, trans 2012)
Critics have compared Tel Aviv-born Keret's stories to works by Kurt Vonnegut, Franz Kafka, even Woody Allen. In the title tale, a group holds the narrator hostage, demanding that he tell them a story.

A Tale of Love and Darkness
by Amos Oz (2002, trans 2004)
This memoir, by one of Israel's best-known writers, chronicles the author's turbulent family life during the State of Israel's early days. It was adapted into a film directed by, and starring, Natalie Portman.

All The Rivers
by Dorit Rabinyan (2014, trans 2017)
Politics gets personal in this novel about an affair between an Israeli woman and a Palestinian man. Israel's Ministry of Education banned this book from the nation's schools.

The Art of Leaving
by Ayelet Tsabari (2019)
Working to reconcile her complicated identity as an Israeli of Yemeni heritage, Tsabari has penned this memoir about her background, her father's death, and her travels around the world.

Watch List

Fill the Void
(2012; drama; directed by Rama Burshtein)
One of the first feature films to be directed by an Orthodox Israeli woman, this movie details life in a Haredi (ultra-Orthodox) community in Tel Aviv.

Lemon Tree
(2008; drama; directed by Eran Riklis)
This story of Israeli-Palestinian relations centres around the Israeli government's decree that a Palestinian woman's lemon trees, located along the Green Line that divides the West Bank from Israel, pose a security threat.

Sand Storm
(2016; drama; directed by Elite Zexer)
If you're curious about contemporary Bedouin society, check out this Arabic-language drama about the lives of two women and their family, set in a small village in southern Israel.

The Band's Visit
(2007; comedy; directed by Eran Kolirin)
Adapted into a Broadway musical that won multiple Tony awards, this comedy tells the tale of an Egyptian band that's unexpectedly stranded in a remote Israeli town.

Waltz with Bashir
(2008; documentary; directed by Ari Folman)
Folman served in the Israeli army during the 1982 war with Lebanon. In this animated film – blending documentary, memoir, even hallucinations – he explores the war's traumatic aftermath.

Eran Riklis recently directed *Spider in the Web*, starring Ben Kingsley and Monica Bellucci.

Born in Jerusalem of Ethiopian descent, Alene, who sings in Amharic, Arabic, English and Hebrew, was set to represent Israel at the Eurovision 2020 finals. Although cancelled, the contest song introduced Alene to a global audience.

The three Haim sisters of A-Wa draw on their Yemeni heritage, combining more traditional music styles with hip-hop and electro beats. This song shares some of the challenges that immigrants from Yemen often face in Israel.

Playlist

Tel Aviv
Omer Adam
Genre: Mizrahi/Pop

Chaim Sheli
Eden Ben Zaken
Genre: Pop

If I Could Go Back in Time
DAM ft Amal Murkus
Genre: Rap

Oof Gozal
Arik Einstein
Genre: Pop

Feker Libi
Eden Alene
Genre: Pop

Hana Mash Hu Al Yaman (Here Is Not Yemen)
A-Wa
Genre: Yemeni/Hip-hop

HaMizrach HaTichon Ivri
Lider
Genre: Pop

Yeroushalayim Shel Zahav
Naomi Shemer
Genre: Ballad

Moonlight
TootArd
Genre: Roots/Dance

Shevet Achim Va'achayot
Various artists
Genre: Pop

Jordan

Read List

Prairies of Fever
By Ibrahim Nasrallah (1985, trans 1993)
Regarded as one of the Arab world's great modernist novels, *Prairies of Fever* weaves a hypnotic tale, infused with lyrical prose, focused on the Arab immigrant experience in the Arabian Peninsula.

Pillars of Salt
By Fadia Faqir (1996)
Themes of repression and the subjugation of female voices are brought to the fore in this novel about two women inmates of a psychiatric hospital set in British Mandate-era Jordan.

The Cat Who Taught Me How to Fly: An Arab Prison Novel
By Hashem Gharaibeh (2010, trans 2017)
Drawing on Gharaibeh's own decade-long experience as a political prisoner in Jordan, this novel lays out the brutality of Jordan's prisons and the powerful roles that hope and dreams play in survival.

Land of No Rain
By Amjad Nasser (2011, trans 2014)
This novel following an exile's return to his Arab homeland, after a plague decimates London, may be futuristic in plot but is woozily poetic in tone.

The Perception of Meaning
By Hisham Bustani (2012, trans 2014)
Bustani is highly regarded for his experimental fiction and this short story collection demonstrates his surreal style. It fires on all cylinders as he pursues themes of technology, violence and modern life.

Watch List

Captain Abu Raed
(2007; drama; directed by Amin Matalqa)
Mistaken for a pilot, an airport janitor is befriended by neighbourhood children and becomes entwined in their lives in this poignant film set in Amman.

Cherkess
(2010; drama; directed by Mohy Quandour)
Set in the Ottoman era, when Circassian refugees were resettled in Jordan, this 'star-crossed lovers' drama focuses on friction between local Bedouin and the new immigrants.

When Monaliza Smiled
(2012; romantic comedy; directed by Fadi G Haddad)
This sweet Amman-set rom-com about the relationship between Jordanian Monaliza and working-class Egyptian immigrant Hamdi highlights Jordanian society's issues with class.

Theeb
(2014; drama; directed by Naji Abu Nowar)
Jordan's critically acclaimed and first Oscar-nominated film follows a Bedouin boy's tale of survival amid Wadi Rum's sumptuously shot desert landscapes during WWI's Arab Revolt.

Tiny Souls
(2019; documentary; directed by Dina Naser)
This documentary is told through the perspective of the Syrian children who live in Jordan's Zaatari refugee camp, the largest Syrian refugee camp in the world.

The rock bridge of Jebel Kharaz in Wadi Rum, the landscape of the film *Theeb*.

Omar Al-Abdallat led the way in converting Jordanian Bedouin folk songs into pop music. Traditionally Bedouin music has three styles: *Taghrud* (camel-caravan songs), *Al-Shi'ir Al-Nabati* (cappella poetry) and *Ayyala* (folk dances).

47 Soul is the leading *shamstep* (Middle Eastern electronic) group; fusing *dabke* beats with EDM and elements of hip-hop. Sham is the Arabic name for Syria, but traditionally it was the name for the entire Levant region.

Playlist

Salma
JadaL
Genre: Rock

Ya Ain Moulayatain
Sakher Hattar
Genre: Classical oud

Ensani Ma Binsak
Diana Karazon
Genre: Arabic pop

Ghazale
Hani Mitwasi
Genre: Traditional Arabic

Drej Ya Ghezaleh
Omar Al-Abdallat
Genre: Modern Bedouin

Mafi Mennik
Aziz Maraka
Jazz-Arabic fusion

Dabke System
47 Soul
Genre: Shamstep

Istanna Shwei
Autostrad
Genre: Rock

Ebn El Nashama
Nedaa Shrara
Genre: Arabic pop

Laykoon
El Morabba3
Genre: Rock

Kenya

Read List

A Grain of Wheat
by Ngũgĩ wa Thiong'o (1967)
One of Africa's greatest voices brings to life a harrowing tale of marriage, imprisonment and exile surrounding the Mau Mau Uprising and Kenya's independence in 1963.

Unbowed
by Wangari Maathai (2006)
The first African woman to win the Nobel Peace Prize, Greenbelt Movement founder Wangari Maathai's uplifting memoir narrates the fraught journey between striving for an education in British Kenya and gaining a position in a chaotic parliament.

Coming to Birth
By Marjorie Oludhe Macgoye (1986)
Intimate coming-of-age narrative following a woman's foray into marriage and motherhood, set alongside the turbulent years of the Kenya Emergency.

The Promised Land
by Grace Ogot (1966)
Ogot examines the relationship between a Tanzanian migrant husband and wife who come face to face with marital and community-driven pressures of gender roles, family and tribal expectations.

Dust
by Yvonne Adhiambo Owuor (2013)
Amid political upheaval in a conflicted nation, a broken family deal with death, grief and an unsettling mystery. The novel spans political unrest from the 1950s up to the violence succeeding Kenya's 2007 presidential elections.

Watch List

Soul Boy
(2010; drama; directed by Hawa Essuman & Tom Tykwer)
Gritty, honest drama about a young man navigating the sprawling Kibera slums when he embarks on a quest to save his ailing father's soul from a female spirit.

Rafiki
(2018; drama/romance; directed by Wanuri Kahiu)
Up against familial pressures and national laws opposing LGBTQ rights, two women fall in love against a vibrant Nairobi. The first Kenyan film to premiere at Cannes Film Festival.

Kati Kati
(2016; drama; directed by Mbithi Masya)
Waking up in the wilderness with no memory, a woman must find a way to understand her place in a sinister afterlife with a group of equally bemused ghosts.

Supa Modo
(2018; drama; directed by Likarion Wainaina)
A young girl diagnosed with a terminal illness dreams of having superhero powers. With a little help from her fellow villagers, she is able to perform miracles in the world around her.

Nairobi Half Life
(2012; drama; directed by David 'Tosh' Gitonga)
An aspiring actor leaves his rural hometown for the capital. Among the bright lights of theatre, he struggles to avoid the looming shadows of the criminal underworld.

Rafiki depicts same-sex relationships in Nairobi (but without any rhinos).

Jambo Bwana's catchy refrains reverberate all over Kenya. After overhearing tourists reciting words in Swahili, Teddy Kalanda Harrison was inspired to write the feelgood global hit that embodies the country's warm, welcoming spirit.

Swinging between Swahili and English, spoken-word and soulful singing, *H_art the Band* is a fashionable trio that dubs its music as 'Afro poetry'. Its upbeat guitar riffs and soft drumbeats have regaled audiences in Nairobi and overseas.

Playlist

Kenge Kenge
Kenge Kenge
Genre: Benga

Nyama Choma
Samba Mapangala & Orchestra Virunga
Genre: Rumba

Dunia Ina Mambo
Eric Wainaina
Genre: Pop

Jambo Bwana
Them Mushrooms
Genre: Reggae/Pop

Dr Binol
Shirati Jazz
Genre: Benga

Ukimwona
H_art the Band
Genre: Pop/Spoken word

Kothbiro
Ayub Ogada
Genre: Traditional Luo

Sema Ng'we
Fena Gitu
Genre: Pop/Hip-hop

Who Can Bwogo Me?
Gidi Gidi Maji Maji
Genre: Hip hop

Sauti
Mercy Myra
Genre: R&B

Lebanon

Read List

Beirut Blues
by Hanan al-Shaykh (1992, trans 1998)
Set amid Lebanon's civil war, this novel is written as a series of letters narrated by a young woman struggling to decide if she should emigrate or stay as her country destroys itself.

Gate of the Sun
By Elias Khoury (1998, trans 2005)
Khoury's epic novel of the Palestinian Nakhba weaves a tale of devastation, survival and hope covering the Palestinian tragedy from the 1948 exodus and through their lives, and liberation struggle, within Lebanon's refugee camps.

June Rain
By Jabbour Douaihy (2006, trans 2014)
A church shoot-out and resulting massacre tears a village in two in a novel which lays bare the sectarian strife and internal divisions that continue to wrack Lebanon.

The Mehlis Report
By Rabee Jaber (2005, trans 2013)
A fictionalised account of the months leading up to the publication of the UN's report on their investigation into Rafik Hariri's assassination, which is also an intimate portrayal of Beirut and its travails.

The Tiller of Waters
By Hoda Barakat (2000, trans 2004)
Narrated by a textile merchant living in his bombed-out shop amid war-wracked Beirut, Barakat's quiet, layered novel widens its scope to take in a deep sweep of history.

Watch List

Safar Barlek
(1966; musical; directed by Henry Barakat)
This musical, starring Fairuz and scored by Lebanon's famed composers the Rahbani brothers, is set in a village during the Ottoman forced-conscription mobilisation in WWI.

West Beirut
(1998; drama; directed by Ziad Doueiri)
A classic of Lebanon's 1990s cinema, this coming-of-age drama revolves around Beirut teen Tarek and his friends as the civil war rages around them.

Where Do We Go Now?
(2011; comedy-drama; directed by Nadine Labaki)
The women of an isolated, mixed-religion village connive against the village men to stop sectarian violence breaking out in this film, which broke Arab cinema's box-office records.

The Insult
(2017; drama; directed by Ziad Doueiri)
Doueiri's Oscar-nominated courtroom drama, centred around a dispute between Lebanese Christian Tony and Palestinian refugee Yasser, explores the wounds left by the civil war.

Capernaum
(2018; drama; directed by Nadine Labaki)
This Oscar-nominated film is a raw depiction of Beirut, following the story of 12-year-old street kid Zain who, while in prison, sues his parents for having him.

Beirut has suffered civil war, sectarian strife and recently a huge explosion, but remains a vibrant coastal city.

Traditionally performed at weddings and celebrations – but just as likely to break out late at night in Beirut's bars – *dabke* is a foot stomping, rhythmic Levantine chain-dance, thought to have origins in Canaanite fertility rituals.

Lebanon's legendary Fairuz transformed the Middle Eastern music scene, moving away from the classical Arabic singing style to create a more modern sound. In her six-decade long career she has sold over 150 million records.

Playlist

Fasateen
Mashrou' Leila
Genre: Rock

Ah W Noss
Nancy Ajram
Genre: Arabic pop

Freedom Denied
Blaakyum
Genre: Folk/Metal

Demoqrati
Fareeq El Atrash
Genre: Hip-hop

El-Tannoura
Faris Karem
Genre: Dabke

Li Beirut
Fairuz
Genre: Lebanese classical

My Mother
Marcel Khalifa
Genre: Classical Arabic

Cliché
Meen
Genre: Rock

Ya Baie
Najwa Karam
Genre: Arabic pop

Herzan
Soap Kills
Genre: Electro-pop

Lesotho

Read List

Chaka
by Thomas Mofolo (1925)
Lesotho-born Mofolo's *King Lear*–like account of Shaka Zulu and the monarch's impact on the region. The first Sesotho book translated into English, it's recognised as one of Africa's greatest 20th-century novels.

She Plays with the Darkness
by Zakes Mda (1995)
The South African novelist returns to mountainous Lesotho, where he came of age in exile under apartheid with his activist father, in this psychological portrait of twins Dikosha and Radisene's fortunes in a remote village.

Everything Lost is Found Again: Four Seasons in Lesotho
by Will McGrath (2018)
The American journalist's account of his year in Lesotho is a love letter to the Mountain Kingdom. It's an empathetic display of African travel writing, which doesn't shy away from tough topics such as AIDS.

The Mountain School
by Greg Alder (2013)
Alder's self-reflective account of his three years with the American Peace Corps in Tsoeneng village, starting with his arrival as the first foreign resident in 37 years.

Singing Away the Hunger: The Autobiography of an African Woman
by Mpho 'M'atsepo Nthunya (1996)
Born into poverty in 1930, Nthunya was a domestic worker for three decades. Her autobiography reveals the hardships of Basotho life, especially for women.

Watch List

This is Not a Burial, It's a Resurrection
(2019; drama; directed by Lemohang Jeremiah Mosese)
Lesotho's first entry for the Best International Feature Film Oscar follows an 80-year-old widow fighting the construction of a dam that threatens to flood her village.

The Forgotten Kingdom
(2013; drama; directed by Andrew Mudge)
The American director spent a year in Lesotho before making this bittersweet rite-of-passage story of a young man returning to his mountain village from Johannesburg to bury his estranged father.

Coming of Age
(2015; documentary; directed by Teboho Edkins)
Edkins grew up in Lesotho and returns for this documentary to follow teenage shepherd brothers Mosaku and Retabile and village girls Senate and Lefa for two years, hearing about their dreams and aspirations.

Dying for Gold
(2018; documentary; directed by Richard Pakleppa & Catherine Meyburgh)
Predominantly shot in Lesotho, this trilingual documentary tells the unheard story of the men who contract silicosis and tuberculosis while digging in the South African gold mines.

Prince Harry in Africa
(2016; documentary; directed by Russ Malkin)
The British royal travels to Lesotho to see progress made by his charity, Sentebale, which provides care, support and education to youth affected by HIV and AIDS.

Lesotho's mountains and canyons feature in the film *The Forgotten Kingdom*.

Morena Leraba is a Basotho shepherd and MC. Also check out his electronic collaborations with Zikomo on *Wena Na* and with the British-Ghanaian group Onipa on *Free Up*, for which he also collaborated with Spoek Mathambo and Syntax.

Typified by accordion, guitar and ululation, Lesotho's homegrown *famo* genre is the country's version of highlife. It originated in *shebeens* (unlicenced bars) frequented by Basotho migrant workers.

Playlist

Lesotho Song
Maseru Band
Genre: Traditional

Khoba – Dance Song
Augustina Mokhosoa
Genre: A cappella

Setsokotsane (Thunder Storm)
African Roots Music ft Lekhotla le Modumo
Genre: Basotho roots

Mzabalazo
RMBO ft Morena Leraba
Genre: Psychedelic

Lefatshe
Puseletso Seema
Genre: Sotho-Tswana folk

Banana Ba Sebokeng
Tau Ea Matsekha
Genre: Famo

Bashanyana
Letsema Matsela & Basotho Dihoba
Genre: Choral

Sadac Lesotho
Apollo Ntabanyane
Genre: Famo

Ho Lokile
Tsepo Tshola
Genre: Gospel

Kingdom King
Various
Genre: Hip-hop

Madagascar

Read List

Beyond the Rice Fields
by Naivo (2012, trans 2017)
The first novel from Madagascar to be translated into English features the intertwined stories of two young children – one enslaved – growing up amid the upheavals and colonial invasions of the 19th century.

Voices from Madagascar
ed by Jacques Bourgeacq & Liliane Ramarosoa (2002)
This dual-language anthology captures the fascinating cultural traditions and Malagasy way of life in beautifully crafted, highly unusual stories. Some pieces have an innate musicality and read more like prose poems.

Return to the Enchanted Island
by Johary Ravaloson (2019)
A complex and engrossing coming-of-age tale about a privileged young Malagasy shipped off to France, and the return to his myth-filled homeland.

The Eighth Continent
by Peter Tyson (2000)
Equal parts natural history, cultural narrative and travelogue, Peter Tyson's impressive volume provides an overview of Madagascar, covering wildlife, spiritual practices and the mysterious origins of its people.

Over the Lip of the World
by Colleen J McElroy (1999)
An American poet and scholar delves into the arcane world of Malagasy oral traditions, meeting storytellers across the country, and sharing their wondrous tales of magic, love and betrayal.

Watch List

Madagasikara
(2018; documentary; directed by Cam Cowan)
Powerful, eye-opening depiction of the systemic reasons behind Madagascar's humanitarian crisis, focusing on three families struggling for survival amid crushing poverty.

Makibefo
(2001; drama; directed by Alexander Abela)
Starring a remote tribe of fishermen, Shakespeare's *Macbeth* is given a bold new interpretation in this mesmerising black-and-white drama about the lust for power.

Fahavalo, Madagascar 1947
(2018; documentary; directed by Marie-Clémence Andriamonta-Paes)
An award-winning documentary about the little-known story of the Malagasy uprising against French colonial rule in 1947, featuring interviews with some of the last survivors from that time.

Quand les Étoiles Rencontrent la Mer
(1996; drama; directed by Raymond Rajaonarivelo)
'*When the Stars Meet the Sea*' is a poetic coming-of-age film about a boy born during a solar eclipse, set against the backdrop of the Malagasy liberation struggle.

Malagasy Mankany
(2012; drama; directed by Haminiaina Ratovoarivony)
Three friends travel from the city to a rural village to help an ailing father in this poignant road movie with touches of adventure, comedy and romance.

Rural life in Madagascar, dominated by poverty, is the subject of *Malagasy Mankany*.

Madagascar's most popular band, Mahaleo has been going since 1972. The band featured in an eponymous 2005 film by Cesar Paes and Raymond Rajaonarivelo, and charismatic frontman Dama even ran for president in 2018.

Traditional Malagasy music revolves around favourite dance rhythms. Energetic *salegy* comes from the Sakalava tribe and has both Indonesian and Kenyan influences. *Tsapika* originated in the south of Madagascar.

Playlist

Mankahery Viavy
Jerry Marcoss
Genre: Salegy

Ikalasoa
Rossy
Genre: Hira gasy

Tovovavy Jefijefy
Rakoto Frah
Genre: Folk

Wouna
Tsew the Kid
Genre: Hip-hop

Somambisamby
Mahaleo
Genre: Folk-pop/Rock

Taniko
Poopy
Genre: Pop

Tia Anao Zaho
Jaojoby
Genre: Salegy

Tolombolagne
Vilon'androy
Genre: Folk

Mifaneva
Rabaza
Genre: Mangaliba

Tsy Vandivandy
Tearano
Genre: Tsapika

Malawi

Read List

For Honour & Other Stories
by Stanley Onjezani Kenani (2011)
Institutions ranging from the UK's Hay Festival to the Caine Prize (the 'African Booker') have acclaimed the short story *Love on Trial* in this Chekhovian collection of tales following Chipiri village's quirky inhabitants.

The Jive Talker: An Artist's Genesis
by Samson Kambalu (2008)
The London-based conceptual artist's coming-of-age memoir of rural Malawi and Kamuzu Academy ('the Eton of Africa') is dominated by his missionary-educated Chewa father, the gregarious jive-talker.

I Will Try
by Legson Kayira (1965)
The late Kayira is one of Malawi's most acclaimed novelists (*The Looming Shadow*, *The Detainee*). This autobiography, covering his journey by foot through the African bush in 1958, was a *New York Times* bestseller.

The Wrath of Napolo
by Steve Chimombo (2000)
Malawi has a strong tradition of poetry; this novel by the late poet Chimombo uses the mythical serpent Napolo as a metaphor for the ills of post-independence Malawi.

The Lower River
by Paul Theroux (2012)
The American travel writer and novelist was stationed in Malawi with the Peace Corps and in this dark novel an American Peace Corps veteran revisits Malawi. Theroux's *Jungle Lovers* and *Dark Star Safari* also cover the country.

Watch List

The Boy Who Harnessed the Wind
(2019; drama; directed by Chiwetel Ejiofor)
The *12 Years a Slave* star directed and acted in this Netflix film about 13-year-old William Kamkwamba, who was inspired by a science book to build a wind turbine, saving his village from famine.

The Boy Who Flies
(2013; documentary; directed by Benjamin Jordan)
This inspirational documentary about Malawi's first paraglider was a film festival favourite for its portrayal of the friendship between a Canadian paraglider and Godfrey, who dreams of flying.

I Am Because We Are
(2008; documentary; directed by Nathan Rissman)
Madonna wrote and narrated this documentary about Malawi's 500,000 HIV and AIDS orphans. It shows her charity Raising Malawi's work with homeless orphans.

William and the Windmill
(2013; documentary; directed by Ben Nabors)
Like *The Boy Who Harnessed the Wind*, this documentary tells the inspirational story of Kamkwamba, who built a power-generating windmill from recycled rubbish.

Child 31
(2012; documentary; directed by Charles Francis Kinnane)
Malawi features in this documentary about the school-dinner charity Mary's Meals, which follows founder Magnus MacFarlane-Barrow's work to uplift millions of children in the developing world.

Playlist

Tsoka Laine
Donald Kachamba's Kwela Band
Genre: Kwela

Blantyre Boma
Erik Paliani
Genre: Jazz/Afrobeats

Born in a Taxi
Blk Sonshine
Genre: Vocal jazz

Zani Muwone
Wambali Mkandawire
Genre: Jazz fusion

This title track from Wambali Mkandawire's debut album garnered the the Mzuzu musician much attention for its mix of jazz and traditional Malawian sounds, leading to more albums and awards.

Warm Heart of Africa
The Very Best
Genre: Afrobeats/Hip-hop

The uplifting *Warm Heart of Africa* song is a collaboration between Mzuzu-born Esau Mwamwaya and London production duo Radioclit, featuring Vampire Weekend's Ezra Koenig. The album of the same name features British songstress M.I.A. on *Rain Dance*, and the group also appears on the Mumford & Sons song *Ngamila* with Senegal's Baaba Maal.

Ndinasangalala
Malawi Mouse Boys
Genre: Gospel

Kasambara
Gwamba
Genre: Chichewa hip-hop

Police Hunt Matafale
Black Missionaries
Genre: Reggae

Amakhala Ku Blantyre
Peter Mawanga & the Amaravi Movement
Genre: Afropop

Chafera Mazira
Danny Kalima
Genre: Afropop

Malawi's lake-filled landscape is the home of Samson Kambalu's charismatic father, *The Jive Talker*.

Mali

Read List

Caught in the Storm
by Seydou Badian Kouyaté (1998)
French-educated Kouyaté was a politician, who wrote the national anthem 'Le Mali', and his novels question the westernisation of Africa. This love story was first published in 1954 as Sous l'orage.

The Timbuktu School for Nomads
by Nicholas Jubber (2016)
The fabled Saharan city of Timbuktu has long cast a spell on the Western imagination. This book describes Jubber's travels with West African nomads as Islamic militants and climate change threaten desert communities.

Dogon: Africa's People of the Cliffs
by Walter E A van Beek & Stephanie Hollyman (2001)
Armchair travel doesn't get much better than this beautifully photographed anthropological study of the cliff-dwelling Dogon people of Bandiagara Escarpment.

Segu
by Maryse Condé (1984)
Set in 1797 in Segou (now part of Mali), this novel by the Guadeloupian winner of the 'alternative Nobel' literature prize follows the Bambara king's chief adviser and his sons, as Islam and slavery threaten their traditional life.

The Bad-Ass Librarians of Timbuktu
by Joshua Hammer (2016)
Hammer gives well-deserved credit to Abdel Kader Haidara and the archivists of Timbuktu, who smuggled 350,000 ancient Arabic texts to southern Mali to save them from destruction by al-Qaeda.

Watch List

Life on Earth
(1998; drama; directed by Abderrahmane Sissako)
'*La Vie sur Terre*' is a bittersweet comedy about a Malian expat returning to Sokolo village. Sissako, a giant of Malian cinema, is the writer, director and lead actor.

The Cultural Journey to Timbuktu
(2015; documentary; directed by Marco Romano)
This documentary shows both the longstanding importance of music to Malians and the impact of the Islamic fundamentalist occupation of northern Mali. Also check out *They Will Have to Kill Us First*.

Mali Blues
(2016; documentary; directed by Lutz Gregor)
A more assured look at Malian music and fundamentalist threats, following four artists and revealing Mali as the birthplace of blues music, which travelled to the US on slave ships.

Timbuktu
(2014; drama; directed by Abderrahmane Sissako)
The Mauritanian-born director's drama about the imposition of Sharia law on northern Mali received many accolades, including a New York Times nomination as one of the 21st-century's best films so far.

Min Ye
(2009; drama; directed by Souleymane Cissé)
Also called '*Tell Me Who You Are*', this drama about adultery in a polygamous marriage is like a Malian *Scenes from a Marriage*. It received a special screening at Cannes.

Near Timbuktu, the nomadic Tuareg people have adapted to life in the desert.

Formed in 1970 as the house band of Bamako's Buffet Hôtel de la Gare, the Rail Band (later known as the Super Rail Band) was the training ground for stars Salif Keïta and Mory Kanté.

Allah Uya is a typically soulful number by the late Grammy-winning guitarist Ali Farka Touré. The Niafunké album on which it appears is named after Touré's home village on the Niger River near Timbuktu. Also look out for his albums with Ry Cooder and kora master Toumani Diabaté. Touré's son Vieux has become a musical star in his own right.

Playlist

Moussolou
Salif Keita
Genre: Afropop

Amassakoul 'N'Ténéré
Tinariwen
Genre: Assouf/Rock

Bi Lambam
Toumani Diabaté with Ballaké Sissoko
Genre: Kora

Sunjata
Super Rail Band
Genre: Afro-Latin

Allah Uya
Ali Farka Touré
Genre: Blues

Slow Jam
Vieux Farka Touré
Genre: Blues

Fadjamou
Oumou Sangaré
Genre: Afropop

M'Bife
Amadou & Mariam
Genre: Afropop

Laidu
Rokia Traoré
Genre: Folk/Traditional

Batoumambe
Habib Koité and Bamada
Genre: Blues

Morocco

Read List

Dreams of Trespass: Tales of a Harem Girlhood
by Fatima Mernissi (1994)
Renowned anthropologist and feminist Fatima Mernissi was born in Fez in 1940, where she grew up in the harem of a *riad* house in the medina. This memoir explores her childhood experiences.

The Last Storytellers
by Richard Hamilton (2011)
Former BBC correspondent Richard Hamilton has recorded the tales told by the few remaining elderly storytellers on Djemaa el-Fna in Marrakesh before they are lost to history.

Secret Son
by Laila Lalami (2009)
A young man brought up in the slums of Casablanca discovers his wealthy father and starts a new life of luxury. But he is soon back on the streets, vulnerable to radicalisation.

The Sand Child (L'Enfant de Sable)
by Tahar Ben Jelloun (1985)
A storyteller in Djemaa el-Fna in Marrakech spins the tale of a father who brings up his eighth daughter as a boy. A lyrical novel of power, colonialism and gender.

For Bread Alone
by Mohamed Choukri (trans by Paul Bowles 1973)
Documenting the harrowing tale of Choukri's family's flight from the Rif Mountains to Tangier at a time of famine, For Bread Alone was once censored in Morocco.

Watch List

Tinghir-Jerusalem: Echoes from the Mellah
(2013; documentary; directed by Kamal Hachkar)
A history of the Berber Jews of Tinghir who, with many other Moroccan Jews, emigrated to Israel due to successful Zionist campaigns in the 1950s.

Marock
(2005; drama; directed by Laïla Marrakchi)
School's out in late-1990s Casablanca, and there's music, fast cars and alcohol. Rita, 17, falls in love with a Jewish boy, falling foul of her place in Muslim culture.

Lalla Aïcha
(2019; drama; directed by Mohamed El Badaoui)
The lives of fishermen are at risk when there's a shortage of fish. Lalla Aïcha is the strong mother who keeps the family together.

The Women in Block J
(2019; drama; directed by Mohamed Nadif)
Three women in a Casablanca psychiatric ward strike up a friendship and slip out at night, finding freedom and hope for the future.

Razzia
(2017; drama; directed by Nabil Ayouch)
Set against a background of social upheaval in Casablanca, the stories of five men and women converge towards a common ideal of freedom.

The Gnaoua are a Sufi brotherhood originating in Mali and Sudan who were brought to Morocco as slaves in the 18th century. Their music, rich with drums and castanets, is used to engender trance to get closer to God.

Playlist

Sigham Olinw
Tasuta N-Imal
Genre: Amazigh Desert Blues

Trap Beldi
Issam
Genre: Rap

Blessed
Shayfeen
Genre: Trap

La Ilaha Ila Lah
Hamid El Kasri
Genre: Gnaoua

Maghadnich Frakek
Cheb Kader
Genre: Rai

360
Manal
Genre: Rap

Inas Inas
Mohamed Rouicha
Genre: Berber folk

Ana Dini Dina Allah
Abderrahim Souiri
Genre: Classical Andalous

Hkayet Lmraya
Saïda Fikri
Genre: Folk/Pop

Sacrifice
MADD
Genre: Rap

Award-winning singer/songwriter Manal may be young, but she's having a profound influence on the Moroccan music industry, where rap is usually a male preserve. She sings in Darija (Moroccan Arabic), French and English.

From snake-charmers to storytellers, you can find almost anything in Marrakesh's Djemaa el-Fna.

Mozambique

Read List

We Killed Mangy-Dog & Other Mozambique Stories
by Luís Bernardo Honwana (1969)
Honwana's influential collection of short stories, a regular on lists of the best 20th-century African literature, explores the struggles of migrant farm workers and other downtrodden Mozambicans near the end of colonialism.

The Tuner of Silences
by Mia Couto (2013)
This book by the best-known and highly acclaimed chronicler of post-colonial Mozambique portrays the intergenerational legacies of war through Mwanito, an 11-year-old boy living in a big-game park.

The First Wife: A Tale of Polygamy
by Paulina Chiziane (2002)
Mozambique's first published female novelist weaves a farce about Rami, who discovers that her policeman husband of 20 years has been leading a double – or rather, a quintuple – life of polygamy.

Ualalapi: Fragments from the End of Empire
by Ungulani Ba Ka Khosa (1987, trans 2017)
Another highly lauded 20th-century literary work, *Ualalapi* tells the story of Mozambique's last Gaza emperor Ngungunhane, debunking the despotic leader's idolisation by post-independence nationalists.

Neighbours: The Story of a Murder
by Lília Momplé (2012)
Mozambican everymen are caught in the tide of history in this novel by veteran writer Momplé, with characters thrown into a conspiracy to destabilise Mozambique.

Watch List

Marrabenta Stories
(2004; documentary; directed by Karen Boswall)
This British-directed documentary tells the Buena Vista All Stars-style story of the veteran stars of Mozambican *marrabenta* music, who formed a band with younger jazz, funk and hip-hop musicians.

The Train of Salt and Sugar
(2016; adventure; directed by Licínio Azevedo)
In this film from Brazilian-born Azevedo, a passenger- and goods-train has to travel 500 miles through rebel territory during the civil war. It was Mozambique's first ever foreign-language film Oscar entry.

Sleepwalking Land
(2007; drama; directed by Teresa Prata)
The Portuguese-Mozambican director took seven years to make this dramatisation of Mia Couto's novel about the civil war, in which an orphaned refugee searches for his mother with an elderly storyteller.

Redemption
(2019; drama; directed by Mickey Fonseca)
The first Mozambican film to feature on Netflix, *Redemption* (aka '*Resgate*') follows well-intentioned ex-con Bruno's descent into Maputo's underworld of kidnapping businesspeople for ransom.

Yvone Kane
(2014; drama; directed by Margarida Cardoso)
Reflecting on Mozambique's brutal civil war and its legacy through female eyes, this film tells an uneasy story of mothers, daughters and the eponymous guerilla fighter.

The influence of Portuguese colonists remains today in Mozambique, as told in *The Tuner of Silences*.

'*Night Bird*' by the marrabenta star Wazimbo featured in a Microsoft advert and a movie, *The Pledge* (2001), directed by Sean Penn and starring Jack Nicholson. The bittersweet song expresses sadness at the promiscuity of a young lady called Maria.

From impoverished Niassa in northern Mozambique, Massukos use their profile to campaign about hygiene, sanitation and HIV/AIDS in rural Mozambique. Guitarist Feliciano dos Santos is the founder of NGO Estamos.

Playlist

A Vasati Va Lomu
Fany Mpfumo
Genre: Marrabenta

Klonipho
Mingas
Genre: Marrabenta

Moral
Lizha James
Genre: Pandza

Majurugenta
Ghorwane
Genre: Afropop/Marrabenta

Nwahulwana
Wazimbo & Orchestra Marrabenta Star de Moçambique
Genre: Marrabenta

Niassa
Massukos
Genre: Afropop

Como Anima a Marrabenta
Neyma
Genre: Kizomba

Too Much ft Kwesta
Laylizzy
Genre: Hip-hop

Story of My Life
Mr Bow
Genre: Jango

Mozambique
Bob Dylan
Genre: Rock

Nigeria

Read List

Things Fall Apart
by Chinua Achebe (1958)
This landmark debut by the late figurehead of African literature is still the continent's most widely read novel. It evokes Igbo village life and the arrival of colonialism and Christian missionaries.

You Must Set Forth at Dawn: A Memoir
by Wole Soyinka (2006)
Soyinka was the first African to receive the Nobel Prize in Literature and this memoir covers his remarkable life of political activism and exile.

Purple Hibiscus
by Chimamanda Ngozi Adichie (2003)
Awarded the Commonwealth Best First Book prize, Adichie's debut tells the coming-of-age story of two privileged teenage siblings from Enugu staying with their aunt, against the backdrop of a military coup.

The Famished Road
by Ben Okri (1991)
Okri became the youngest ever Booker Prize winner for this novel, a singular mix of magical realism and Yoruba folklore. Protagonist Azaro, an *abiku* or spirit child, is pulled between the material and spirit worlds.

The Joys of Motherhood
by Buchi Emecheta (1979)
Emecheta's novel covers the common theme of tension between patriarchal African traditions and westernisation, seen through the hardships experienced during WWII by Ona and her daughter Nnu Ego.

Watch List

Lionheart
(2018; drama; directed by Genevieve Nnaji)
Nnaji's directorial debut was both the first Netflix original film produced in Nigeria and the country's first submission to the Oscars. She plays a businesswoman tackling a male-dominated industry.

Omo Ghetto: The Saga
(2020; comedy; directed by Funke Akindele & J J C Skillz)
In this gangster-comedy sequel, the highest-grossing Nollywood movie to date, co-director Akindele also stars as ghetto girl Lefty, who is adopted to live with her long-lost twin sister.

The Wedding Party
(2016; romantic comedy; directed by Kemi Adetiba)
Another record Nollywood earner, this female-directed rom-com shows a couple's lavish wedding descending into a nightmare of exes, squabbling parents and uninvited guests.

October 1
(2014; thriller; directed by Kunle Afolayan)
As colonial Nigeria prepares for independence from the British in 1960, a police officer hunts a serial killer murdering young women in remote western Akote.

Ghetto Dreamz: The Dagrin Story
(2011; documentary; directed by Daniel Ademinokan)
This documentary about the tragically short-lived rap star Dagrin, who died aged 25 in a car crash, shows the challenges faced by young Nigerian musicians.

The Wedding Party director Kemi Adetiba studied at the New York Film Academy.

Fela Kuti is Nigeria's pidgin-singing king of Afrobeat, who had 27 'queens' and established an independent republic (watch *Finding Fela!* to learn more). It's hard to single out just one of his tracks for special mention. This 1989 song is typically political and long at over 20 minutes.

Apart from Kuti, Grammy-nominated Adé is Nigeria's biggest global name. Based on a traditional African proverb, this song from his international breakthrough album *Juju Music* (1982) means 'fight for me' in Yoruba.

Playlist

Atewo-Lara Ka Tepa Mo 'Se
Segun Adewale
Genre: Yo-pop

Tuo De Miyan
IK Dairo and His Blue Spots
Genre: Jùjú

Truth Don Die
Femi Kuti
Genre: Afrobeat

I Come From De Ghetto
Majek Fashek
Genre: Reggae

Beasts of No Nation
Fela Kuti
Genre: Afrobeat

Ja Funmi
King Sunny Adé & His African Beats
Genre: Afropop/Jùjú

Ma Lo
Tiwa Savage ft Wizkid
Genre: Afropop

Green Light
Olamide
Genre: Afropop/Hip-hop

Never (Lagos Never Gonna Be the Same)
Tony Allen & Hugh Masekela
Genre: Afrobeat/Jazz

Kanayo
Eva Alordiah
Genre: Hip-hop

Saudi Arabia

Read List

Cities of Salt
By Abdelrahman Munif (1984, trans 1987)
The discovery of oil in an unnamed Arabian Gulf country devastates a traditional Bedouin community in this epic novel from Saudi-Jordanian Munif. A classic of 20th-century Arab literature.

Adama
By Turki al-Hamad (1998, trans 2003)
This coming-of-age story of a Saudi teenager's political awakening amid the turbulent 1960s and '70s managed to become a bestseller in the Middle East, despite being banned by several countries.

Wolves of the Crescent Moon
By Yousef al-Mohaimeed (2003, trans 2007)
Life on the fringes of modern Saudi society are laid bare in this short, Riyadh-set novel that interweaves the stories of three men scraping by beneath the city's glitzy façade.

Girls of Riyadh
By Rajaa Alsania (2005, trans 2008)
Four young women look for love in this breezy, page-turning novel that gained the label 'Saudi chick-lit', but also offers insights into the daily lives of Saudi women.

The Dove's Necklace
By Raja Alem (2010, trans 2016)
Both dark crime-mystery and portrait of Mecca's underbelly, Alem's novel, which won the International Prize for Arabic Fiction, opens with a naked female body found in a Meccan alleyway.

Watch List

Wadjda
(2012; drama; directed by Haifaa al-Mansour)
A young girl enters a Quranic-recitation contest in the hope of winning the prize money to buy a bicycle in al-Mansour's heartwarming and critically acclaimed debut.

Barakah Meets Barakah
(2016; romantic comedy; directed by Mahmoud Sabbagh)
Sabbagh's Jeddah-set film follows the comic trials of Barakah and Bibi as they attempt to conduct a romance amid Saudi's social restrictions.

Wasati
(2016; drama; directed by Ali Kalthami)
This short-feature drama is based on the real-life extremist attack on a Riyadh theatre while it was staging the play *Wasati bela Wastiah*.

One Day in the Haram
(2017; documentary; directed by Abrar Hussain)
This sumptuously shot documentary, covering a 24-hour period inside Mecca's Masjid al-Haram, is the first time cameras have been given unrestricted access to Islam's holiest place.

The Perfect Candidate
(2019; drama; directed by Haifaa al-Mansour)
In al-Mansour's gentle second drama, one town's local elections shine a spotlight on conservative societal mores when a female doctor decides to run for office.

Saudi Arabia's most famous folk dance is the *ardah*. Originally a war-dance, the *ardah* is a slow-paced line-dance, performed with swords and accompanied by rhythmic drumming and the singing of poetry.

Afro-Arab singer Etab was Saudi's first female performer. A star in the 1960s and '70s, her concerts were considered risqué and the Saudi authorities' disapproval contributed to her move to Egypt in the late '70s.

Riyadh is the setting for Ali Kalthami's true-life account of terrorism in the movie *Wasati*.

Playlist

Ashofak Kil Youm
Mohammed Abdu
Genre: Classical Arabic

Ride the Wave
Omar Basaad
Genre: Electronic

Umm El Dunia
Qusai ft Sadat & Fifty
Genre: Hip-hop

Samri
Saqa Sowab Al-Haya
Genre: Nadj-region folk

Ardah
Folkloric troupes throughout the country
Genre: Folk dance

Gani El Asmar
Etab
Genre: Khaleeji

Marrat
Talal Maddah
Genre: Classical Arabic

Pinocchio
The AccoLade
Genre: Rock

La Ya Habibi
Ibtisam Lutfi
Genre: Khaleeji

Atba'a Al-Namrood
Al-Namrood
Genre: Folk-metal

Senegal

Read List

At Night All Blood is Black
by David Diop (2018)
This acclaimed novel of bullets and black magic tells the forgotten story of Senegalese soldiers in the WWI trenches, focusing on Alfa's descent into madness on night visits to No Man's Land.

Tales of Amadou Koumba
by Birago Diop (1947, trans 1966)
In this classic folktale collection, Diop transcribed the tales he heard from a Senegalese village *griot* (storyteller), providing a fascinating insight into the beliefs of West Africa almost a century ago.

So Long A Letter
by Mariama Bâ (1981)
This novel by the late feminist writer Mariama Bâ expresses frustration about women's position in polygamous society, telling the protagonist's story through a letter to a friend in the US.

God's Bits of Wood
by Ousmane Sembène (1962)
Examining the Senegalese and Malian experience of French colonialism, this fictionalised account describes a strike on the Dakar–Niger railway.

The Music in My Head
by Mark Hudson (1998)
Latterly an art critic for the UK's *Daily Telegraph*, Hudson's satire of the world music scene follows washed-up record-label man Andrew 'Litch' Litchfield to N'Galam (Dakar) in search of the next big thing.

Watch List

Griot
(2011; documentary; directed by Volker Goetze)
A German director's examination of the *griot* tradition, whereby music is passed on as a birthright in certain families, through the story of *kora* player Ablaye Cissoko.

Aujourd'hui
(2012; drama; directed by Alain Gomis)
This story from a Senegalese director is set against the bustling backdrop of Dakar, where lead protagonist Satché seizes the day in the face of impending death.

La Pirogue
(2012; drama; directed by Moussa Touré)
This topical story focuses on a Senegalese fisherman smuggling people across the Atlantic to Spain in a pirogue. The film's Senagalese star, Souleymane Seye Ndiaye, went on to appear in French film *Montparnasse Bienvenüe* and the Sky series *Zero Zero Zero*.

Journey of the Hyena
(1973; drama; directed by Djibril Diop Mambéty)
This road movie is a Senegalese *Easy Rider*: a young couple dream of escaping Dakar riding a motorbike with a mounted bull-horned skull.

Moolaadé
(2004; drama; directed by Ousmane Sembène)
The late Senegalese cinema giant was the first African director to gain international recognition. Critics including Roger Ebert praised Sembène's final film, which tackles the issue of female genital mutilation.

Watch *Aujourd'hui* to experience Senegal's coastal capital, Dakar.

N'Dour was an African superstar even before this global hit of 1994, rising through the bands of '70s and '80s Dakar that developed *mbalax* – a danceable mix of Wolof *sabar* percussion with Cuban and Latin American rhythms.

Hailing from a *griot* family from Ziguinchor in southwest Senegal, Keita is now based in the UK. His mastery of the 21-stringed *kora* finds lyrical expression in this duet with celebrated Welsh harpist Finch. It's one of Keita's many crossover collaborations; another is Damon Albarn's *Africa Express*.

Playlist

Traveller
Baaba Maal
Genre: World beat

Utrus Horas
Orchestra Baobab
Genre: Son Cubano/Wolof

Seyni
Cheikh Lô
Genre: Afro-Cuban/Mbalax

7 Seconds
Youssou N'Dour & Neneh Cherry
Genre: Mbalax

Future Strings
Seckou Keita & Catrin Finch
Genre: Classical strings

Milyamba
Sister Fa
Genre: Hip-hop

Jalo
Étoile de Dakar ft Youssou N'Dour
Genre: Mbalax

Sey
Africando All Stars ft Thione Seck
Genre: Afro-salsa

Issoire
Nuru Kane
Genre: Gnawa/Rock

Nice
Xuman
Genre: Hip-hop

Sierra Leone

Read List

Ancestor Stones
by Aminatta Forna (2006)
Deftly combining personal and political with a story spanning five lives and the entire 20th century, this is one of many critically acclaimed works by Sierra Leone's foremost living novelist.

A Long Way Gone
by Ishmael Beah (2007)
The memoir of one of many thousands of child soldiers forced into combat during Sierra Leone's 10-year civil war, this unflinchingly raw account is ultimately a tale of redemption.

The Palm Oil Stain
by Nadia Maddy (2011)
A touching love story set against the backdrop of civil war, Shalimar escapes an attack on her village and forges an unlikely bond with her rescuer.

Sacred River
by Syl Cheney-Coker (2013)
Replete with poetic language and peppered with magic realism, this novel centres on a legendary 19th-century emperor reincarnated as the head of a fictionalised West African nation.

No Past, No Present, No Future
by Yulisa Amadu Maddy (1973)
A bildungsroman charting the lives of three young men – including, controversially for the time, a gay character – and their evolving friendship against the backdrop of colonial Africa.

Watch List

Process
(2017; music; directed by Kahlil Joseph)
With cinematic visuals from the man behind Beyonce's *Lemonade* video, this dream-like tribute to the Freetown roots of R&B musician Sampha is narrated by his nonagenarian grandmother.

Sierra Leone: A Culture of Silence
(2014; documentary; directed by Raouf J Jacob)
Thirteen years after fleeing the civil war, the filmmaker returns to his home country where he explores sensitive cultural issues including FGM and the mining of so-called 'blood diamonds'.

Sierra Leone's Refugee All Stars
(2005; documentary; directed by Zach Niles & Banker White)
The uplifting story of how a group of singers, drummers and guitarists – displaced by the 1991-2002 civil war – were brought together and offered hope by music.

Amistad
(1997; drama; directed by Steven Spielberg)
Based on real events, enslaved Mende tribesmen en route to the US in 1869 gain control of their captors' ship – but must then battle the courts for their freedom.

Survivors
(2018; documentary; directed by Anna Fitch, Lansana Mansaray, Arthur Pratt & Banker White)
A startlingly intimate portrait of the Ebola epidemic, showing its devastating impact and the extraordinary resilience and courage of those at the frontline.

Author Aminatta Forna's most recent essay collection is *The Window Seat: Notes From A Life In Motion.*

Maringa, or 'palm wine music', is named after the preferred beverage of its early protagonists – Kru musicians, who played guitars brought by Portuguese sailors to create a distinctive style combining local melodies and Caribbean-inspired calypso.

Meaning 'potbellied boy', the 2015 track *Borbor Bele* is one of Emmerson Amidu Bockarie's many anti-corruption protest songs. Singing in English and Krio, he's caused a stir with his explicit criticism of Sierra Leone's ruling elite.

Playlist

Let me love you
Bunny Mack
Genre: Funk

Ajuba
Dr Oloh
Genre: Gumbe

Diyara
Fantacee Wiz
Genre: Folk

E Sidom Panam
African Connexion
Genre: Afropop

Lumley
Ebenezer Calendar & His Maringa Band
Genre: Maringa

Kam Join We Bo
Yung Sal
Genre: Hip-hop

Borbor Bele
Emmerson
Genre: Afropop

Please Go Easy With Me
S E Rogie
Genre: Maringa

Around Ya
Star Zee
Genre: Dancehall

Shukubly
Drizilik
Genre: Rap

South Africa

Read List

Long Walk to Freedom
by Nelson Mandela (1994)
The 1993 Nobel Peace Prize winner and former president of South Africa, Nelson Mandela chronicles his lifetime fighting racial injustice and his 1990 release from over 25 years of political imprisonment.

Coconut
by Kopano Matlwa (2007)
In this moving debut novel, Fiks and Ofilwa struggle to find their identities as black women growing up in the all-white environment of suburban South Africa.

Disgrace
by J M Coetzee (1999)
The 2003 Nobel Prize in Literature winner Coetzee delves into the mind of David Lurie, a twice-divorced professor who loses his job, and the existential crisis that unravels his life thereafter.

The Conservationist
by Nadine Gordimer (1974)
Joint winner of the 1974 Booker Prize, *The Conservationist* follows a wealthy white businessman in apartheid South Africa as he is presented with a moral dilemma when he finds a corpse on his property.

Born a Crime: Stories from a South African Childhood
by Trevor Noah (2016)
This comedic autobiography from *The Daily Show* host Noah explores his adolescence and subsequent rise to fame as a biracial boy born during apartheid South Africa, when interracial marriage was illegal.

Watch List

The Wound
(2017; drama; directed by John Trengove)
This award-winning film shares the story of two closeted gay men through the lens of the Xhosa rite of Ulwaluko—a ritual preparing boys for manhood.

Matwetwe (Wizard)
(2011; comedy; directed by Kagiso Lediga)
A coming-of-age tale about Lefa and Papi, two best friends and recent high-school graduates who embark on the adventure of their lives one New Year's Eve in a South African township.

Die Seemeeu (The Seagull)
(2018; drama; directed by Christiaan Olwagen)
Anton Checkov's famous play *The Seagull* is adapted for the screen in this film, which takes place in 1990s South Africa 100 years after it was originally written. The film is in Afrikaans with English subtitles.

Moffie
(2019; drama; directed by Oliver Hermanus)
Named for the South African slur for gay men, this emotive film follows a young gay man in 1981 South Africa as he completes his two years of military service while witnessing racism, homophobia and brutality.

Five Fingers for Marseille
(2018; Western; directed by Michael Matthews)
A classic Western set in the savannah of post-apartheid South Africa, about a man who flees police aggression in his hometown – a place called Marseille – but then later returns to protect it from a new threat.

Singer-songwriter Vusi Mahlasela played a key role in the anti-apartheid movement.

Vusi Mahlasela's signature voice and sound is lauded as the unofficial 'Voice of South Africa.' The reigning king of contemporary African folk, he has released seven top-selling studio albums and is a multigenerational favourite.

Black Coffee has established himself as the number one South African DJ and one of the top DJs internationally. His catchy synth-pop hits show a masterful mix of catchy lyrics, turntable skills and epic guest appearances.

Playlist

John Cena
Sho Madjozi
Genre: Pop/Hip-hop

Boss Zonke
Riky Rick
Genre: Hip-hop

Daylight
Muzi
Genre: Pop

Impi
Johnny Clegg
Genre: Pop

Say Africa
Vusi Mahlasela
Genre: Folk

Weekend Special
Brenda Fassie
Genre: Pop

Never Gonna Forget
Black Coffee ft Diplo
Genre: Electronic/Pop

Echo of the Sun
Aidan Martin
Genre: Rock

Thati Sgubu
Bongo Maffin
Genre: Pop

Syria

Read List

Fragments of Memory
By Hanna Mina (1975, trans 1993)
Set in the 1930s and 1940s, Mina's slow-paced novel documents the lives of a poverty stricken Syrian family trying to scrape a living during French rule.

Sabriya: Damascus Bitter Sweet
By Ulfat Idilbi (1980, trans 1995)
Narrated by a young woman in 1920s Damascus, this novel brings alive the era's suffocating atmosphere of societal restrictions for females. Its backdrop is the nationalist uprising against the French.

The Dark Side of Love
By Rafik Schami (2004, trans 2009)
A multigenerational cast of characters deal with local blood feuds, colonial oppression, love-rivalry, national uprisings, coups and sectarian violence in this sweeping Damascus-set saga that traverses Syria's 20th century.

No Knives in the Kitchens of this City
By Khaled Khalifa (2013, trans 2016)
This portrayal of one Aleppo family's struggles and eventual slide into catastrophe through the decades of Hafez al-Assad's dictatorship is Khalifa's most brutal and haunting work.

The Frightened Ones
By Dima Wannous (2017, trans 2020)
The crushing personal psychological toll of violence, loss and exile centre this novel, which chronicles Syria's years of dictatorship and the road to revolution and civil war.

Watch List

Dreams of the City
(1984; drama; directed by Mohamed Malas)
This coming-of-age tale follows Deeb and his family as they seek a new life in Damascus, set against the backdrop of Syria's politically turbulent 1950s.

The Night
(1992; drama; directed by Mohamed Malas)
Malas' film focused on his hometown Quineitra (which was captured by Israel during the 1967 war), and explores the Arab-Israeli conflict through Syrian eyes.

Damascus with Love
(2010; drama; directed by Mohamad Abdulaziz)
A love-letter to Damascus' twisty lanes, crumbling architecture and hidden corners that follows Jewish-Syrian Hala's search for the Christian-Syrian love interest she had previously thought dead.

Ladder to Damascus
(2013; drama; directed by Mohamed Malas)
Shot in secrecy, this drama from Malas weaves the tales of a group of artistic young Syrians amid the revolution, whose lives are slowly consumed by war.

For Sama
(2019; documentary; directed by Waad al-Kateab)
Al-Kateab filmed her life in Aleppo for five years following the city's uprising, resulting in this searing documentary that captures a woman's experience of conflict.

Author Khaled Khalifa won the Naguib Mahfouz Medal for Literature in 2013.

Always in his trademark sunnies and red-and-white chequered *keffiyeh*, Omar Souleyman is Syria's iconic *dabke* star. He's brought his electronic *dabke* beats and vocals to remixes for Björk, and played Glastonbury and the Roskilde festival.

Like many Syrian refugees, indie-rockers Khebez Dawle took the perilous route across the Mediterranean into Europe. Along the way they promoted their music, giving CDs to tourists when they landed on the beach in Lesbos.

Playlist

Fouq Annakhl
Sabah Fahkri
Genre: Classical Arabic

Lama Bada Yatathana
Lena Chamamyan
Genre: Syrian folk

Zamilou
Bu Kholthoum
Genre: Hip-hop/R&B

Kalam Ennas
George Wassouf
Genre: Classical Arabic

Warni Warni
Omar Souleyman
Genre: Dabke

Belsharea'
Khebez Dawle
Genre: Alt rock

Tarab Dub
Hello Psychaleppo
Genre: Arabic trip-hop

Fight or Flight
Chyno With a Why?
Genre: Hip-hop

Ya Teira Tiri
Zein Al-Jundi
Genre: Syrian folk

Oghneyat Men Baladi
The Orchestra of Syrian Musicians
Genre: Classical Arabic

Togo

Read List

The Shadow of Things to Come
by Kossi Efoui (2011, trans 2013)
Efoui, a prolific writer, was an exile of the late Gnassingbé Eyadéma's autocratic presidency in Togo. In this novel he explores the 'resource curse' and the exploitation of oil in an unnamed African nation.

An African in Greenland
by Tété-Michel Kpomassie (1981)
One of the few Togolese titles available in English evokes not the steamy palm forests of West Africa, but the icy Arctic Circle. Kpomassie worked his way north to adventure among the Inuit.

The Village of Waiting
by George Packer (1984)
Adding to the body of African travelogues by American Peace Corps veterans, Packer recounts his experiences as a *yovo* (foreigner) in the village of Lavié during Gnassingbé Eyadéma's years of rule.

Do They Hear You When You Cry?
by Fauziya Kassindja (1999)
Kassindja's account of being forced into an arranged marriage at 17 and told to prepare for *kakia* (female genital mutilation) is a harrowing story, ending with her landmark battle for US asylum.

The Fixer: Visa Lottery Chronicles
by Charles Piot (2019)
An anthropology professor follows visa-fixer Kodjo Nicolas Batema and his clients at the US Embassy in Togo, obsessed with the idea of making it to the US.

Watch List

Kondonna
(2011; documentary; directed by Luc Abaki Kouméabalo)
A Togolese journalist's documentary about the traditional coming-of-age ceremonies that initiate young men into adulthood, seen from the critical but sympathetic perspective of an educated observer.

Ashakara
(1991; drama; directed by Gérard Louvin)
Togo-set medical drama about the fight between a doctor who's manufacturing the medicine to treat a deadly virus in Africa, and a rival pharmaceutical giant.

T'bool
(2019; documentary; directed by Joel M'Maka Tchédré)
One of a few documentaries by this Togolese director, *T'bool* covers the sacred 'fire dance' ceremony of northern Togo.

Women with Eyes Open
(1994; documentary; directed by Anne-Laure Folly)
The Color Purple author Alice Walker praised this Togolese film, '*Femmes aux Yeux Ouverts*', which shows women from across West Africa talking about the repression they face.

La Vie de Daniel
(2018; drama; directed by Gilbert Bararmna)
This gritty short by a young Togolese director focuses on the tensions between an autistic man and his half-sister, who struggles to accept him.

Fally Ipupa's 2006 debut album, *Droit Chemin*, sold over 100,000 copies

Bella Bellow is a legend of modern Togolese music, and could have been Francophone Africa's version of Miriam Makeba were it not for her untimely death in a car crash in 1973. Singing sweetly in French and Ewe, her '60s-pop recordings in Paris with Togolese producer Gérard Akueson and Cameroonian saxophonist Manu Dibango influenced stars such as Benin's Angélique Kidjo.

Accompanied here by Congolese *ndombolo* star Fally Ipupa, Toofan (aka Master Just and Barabas) invented Cool-Catché – an African dance craze.

Playlist

Mifon
King Mensah
Genre: Afropop/Ewe/Kabye

Blok Chômage
Peter Solo
Genre: Afro-funk

Chapeau
Caline Georgette ft Jimi Hope
Genre: Rock

G Blem Di
Akofa Akoussah
Genre: Afro/Folk/Funk

Avoudé
Prince Mo
Genre: Hip-hop

Rockia
Bella Bellow
Genre: Afro-soul

Saka Saka
Almok
Genre: Afropop/Hip-hop

Yé Mama
Toofan ft Fally Ipupa
Genre: Afropop/Hip-hop

Bouger Bouger
Nimon Toki Lala
Genre: Soukous

Adome Nyueto
Yta Jourias
Genre: Soul

Tunisia

Read List

Muqaddimah
by Ibn Khaldun (1377, translated 1958)
One of Islam's greatest philosophers, Tunisian Ibn Khaldun is considered the father of historiography, sociology and economics. This work explores the history of empires and his analysis of social cohesion.

Return to Dar al-Basha
by Hassan Nasr (1994, translated 2006)
After a traumatic childhood, Murtada returns to the old city of Tunis where the sights, sounds and smells spark his memories, leaving him contemplating the past in a changing society.

A Tunisian Tale
by Hassouna Al-Mosbahi (2007, translated 2011)
Seen from the perspective of a young man on Death Row and from that of his mother speaking from beyond the grave, this novel confronts the boundaries of Tunisian society.

Behind Closed Doors: Tales of Tunisian Women
by Monia Hejaiej (1996)
The tradition of women telling stories at community events goes back a long way in Tunisia. Ghaya, Sa'diyya and Kheira spin tales of love, relationships, morality and the lot of women.

The Present Tense of the World: Poems 2000-2009
by Amina Saïd (2011)
One of Tunisia's leading poets, Saïd has a mixed French and Tunisian heritage that informs her work. In this collection, she explores questions of identity and place.

Watch List

Dachra
(2018; horror; directed by Abdelhamid Bouchnak)
In this rare Tunisian horror film – Bouchnak's first – three students visit a strange town to investigate terrifying occult rituals. Based on a true story.

As I Open My Eyes
(2015; drama; directed by Leyla Bouzid)
Hailed as one of the best films about the Arab Spring, this story features a teenage rock star who gets caught up in politics.

The Flower of Aleppo
(2016; drama; directed by Ridha Behi)
A broken marriage, political upheaval and a beloved son ensnared by ISIS threaten to overwhelm Salma. She penetrates war-ravaged Aleppo to save her son.

Noura's Dream
(2019; drama; directed by Hinde Boujemaa)
With her husband in jail, Noura starts divorce proceedings so she can marry her lover. But her husband is released early, and the penalty for adultery is harsh.

Look at Me
(2018; drama; directed by Nejib Belkadhi)
Leading a low life in Marseilles, Lotfi must return to Tunisia when his estranged wife has a stroke, to take care of his autistic son.

Discover the old quarters of Tunis in Hassan Nasr's *Return to Dar al-Basha*.

The Gnaoua music of Tunisia, *stambeli* arrived with slaves from sub-Saharan Africa in the 18th century. The music, dance and chants are both a healing and religious practice where performers go into a trance and are said to embody supernatural entities.

Influenced by Joan Baez, Björk and Massive Attack, this ballad became the anthem of the Arab Spring. Mathlouthi performed it in Oslo at the 2015 Nobel Peace Prize ceremony when the Tunisian National Dialogue Quartet won the award.

Playlist

Ibtihal
Lotfi Bouchnak
Genre: Traditional Malouf

Diwan of Beauty and Odd
Dhafer Youssef
Genre: Jazz

Jadakal Ghaythu
Sonia M'Barek
Genre: Classical Arab

Ya Lili Ya Lila
Balti
Genre: Rap

Boussadia
Dendri Stambeli Movement
Genre: Stambeli

Kelmti Horra
Emel Mathlouthi
Genre: Ballad/Protest

3lech Nloum
Anis Rouid
Genre: Traditional mizwad (bagpipes)

Gangsta
Medusa TN
Genre: Hip-hop/Rap

Born to Survive
Myrath
Genre: Progressive metal

Me Dayem Welou
Samara
Genre: Rap

Turkey

Read List

The Time Regulation Institute
by Ahmet Hamdi Tanpınar (1962, trans 2014)
Tanpıner's allegory mocking Turkey's radical modernisation push is a masterpiece, set in a surreal Istanbul where the Time Regulation Institute dictates that all clocks be set to Western time.

Memed, My Hawk
by Yaşar Kemal (1955, trans 1961)
A lyrical evocation of the harsh scrabble of rural life in Kemal's native Çukurova region, this beloved classic follows Memed from poverty wracked childhood through transformation into heroic outlaw.

Dear Shameless Death
by Latife Tekin (1983, trans 2001)
Sprinkled with encounters with *djinn* and folkloric entities, Teken's magical-realism tale of a village family's move to the city tackles the effects of swift modernisation and Turkey's mass rural-to-urban migration shift.

Snow
by Orhan Pamuk (2002, trans 2004)
Set in a snowbound Kars, a powder-keg is ignited after a female student's suicide. This fizzling novel from Nobel laureate Pamuk tackles Turkey's 20th-century clashes between tradition, religion and modernity.

10 Minutes 38 Seconds in this Strange World
by Elif Şafak (2019)
Turkish-British author Şafak made the Booker Prize shortlist for this novel about the memories of a murdered sex-worker in Istanbul, in the last minutes of her life.

Watch List

Yol
(1982, drama; directed by Yılmaz Güney & Şerif Gören)
This bleak portrayal of life under Turkey's military dictatorship, following the stories of five prisoners on home-leave, won the Palme d'Or at Cannes.

Vizontele
(2001; comedy-drama; directed by Yılmaz Erdoğan & Ömer Faruk Sorak)
Television arrives in a rural village in Turkey's far southeast and a tug-of-war between welcoming and oppositional factions ensues in writer-director Erdoğan's beloved film.

Once Upon a Time in Anatolia
(2011; drama; directed by Nuri Bilge Ceylan)
Steeped in a brooding atmosphere, this slow-paced film about the hunt to exhume the body of a murdered man won Ceylan the Grand Prix at Cannes.

Winter Sleep
(2014; drama; directed by Nuri Bilge Ceylan)
Ceylan returned to Cannes and won the Palme d'Or for this slow-burn epic about the powerful and powerless, set amid the glowering, wintry landscapes of Cappadocia.

Mustang
(2015; drama; directed by Deniz Gamze Ergüven)
This debut film by Turkish-French Ergüven follows the fates of five sisters in a Black Sea village as they rebel against the restrictions placed upon them.

Playlist

Şımarık
Tarkan
Genre: Turkish pop

Yemenimde Hare Var
Turan Saka & Mustafa Ertürk
Genre: Fasıl

Yoluna Yoldaş Olam
Aynur Doğan
Genre: Contemporary Kurdish

Çığrık
Moğollar
Genre: Anatolian rock

From the 1960s through to the early '80s, Turkish musicians began fusing western-style psychedelic rock with Turkish folk music. The band Moğollar were one of the pioneers of this scene, which became known as Anadolu (Anatolian) rock.

Seni Sana Bırakmam
İbrahim Tatlises
Genre: Arabesk

The Sema Ritual
Galata Mevlevi Sema Ensemble
Genre: Religious

The Mevlevi Sema is the whirling dervish religious ceremony. The ritual's hypnotic music is traditionally provided by the ney flute, kudüm drum and oud, and accompanied by the chanting of Koranic verses.

Hadi Bakalım
Sezen Aksu
Genre: Turkish pop

Aşıkların Sözü Kalır
Baba Zula
Genre: Indie

Cevapsız Sorular
Manga
Genre: Rock

Susamam
Şanışer
Genre: Hip-hop

Cappadocia's strange landscape is the location for the story of *Winter Sleep*.

United Arab Emirates

Read List

The Diesel
By Thani Al-Suwaidi (1994, trans 2012)
Al-Suwaidi's surreal and poetic short novel, narrated by a transgender character, is steeped in themes of identity and transition – both personal and societal.

The Sand Fish
By Maha Gargash (1999, trans 2009)
Set in the 1950s, this Dubai love story provides a vivid portrayal of the harsh, simple life most Emiratis lived in the years before the oil boom.

In a Fertile Desert: Modern Writing from the United Arab Emirates
By Denys Johnson-Davies (2009)
As a theme, the UAE's fast-paced transformation from poverty-stricken backwater to one of the wealthiest corners of the world percolates throughout this short-story collection from Emirati writers.

Temporary People
By Deepak Unnikrishnan (2017)
Infused with magical realism, this novel of interlinked stories explores the fates of the UAE's foreign nationals, spotlighting the lives of the people who make up over 80% of the country's population.

Gathering the Tide: An Anthology of Contemporary Arabian Gulf Poetry
Edited by Jeff Lodge, Patty Paine & Samia Touati (2011)
This collection of poetry spans works from across the Arabian Peninsula (some published in English here for the first time), including many entries from Emirati poets.

Watch List

City of Life
(2009; drama; directed by Ali F Mostafa)
Several storylines following different lives in Dubai weave together and interlink in this ode to the city's multicultural population.

Sea Shadow
(2011; drama; directed by Nawaf Al-Janahi)
A gentle, coming-of-age drama about two Emirati teens navigating the boundaries of societal traditions, family commitments and religious values as they grow up.

Bilal: A New Breed of Hero
(2015; animation; directed by Khurram H Alavi & Ayman Jamal)
This 3D computer-animated film depicting the life of Bilal ibn Rabah (one of the first Muslims) is the first full-length animated feature produced in the Middle East.

Going to Heaven
(2015; drama; directed by Saeed Salmeen Al-Murry)
The importance of family in Emirati life centres this drama about a young boy's journey from Abu Dhabi to Fujairah, to find his estranged grandmother.

Shabab Sheyab (On Borrowed Time)
(2018; comedy-drama; directed by Yasir Al-Yasiri)
Four grumpy, elderly friends break out of their Dubai retirement home and head into the city for adventure, in this utterly charming, warm comedy.

Around Dubai's creek with its *abras* (river boats), the city retains its original way of life.

Dubai-based death-metal band Nervecell are one of the pioneers of UAE metal. Formed in 2000, the band has released three albums and toured across Asia and Europe, gaining wide recognition within the global metal scene.

Emirati traditional folk dances include the *Ayyalah*, a slow-paced, rhythmic male line-dance performed with camel sticks, and the *Liwa*, a circle-dance with its roots in East Africa that uses the *mizmar* (wind instrument) and drums.

Playlist

Caravan
Kamal Musallam
Genre: Jazz-Arabic fusion

Falling
Hamdan Al-Abri
Genre: Contemporary R&B

Philosophies
Abri
Genre: Soul/Pop

Upon an Epidemic Scheme
Nervecell
Genre: Metal

Easy
Esther Eden
Genre: Pop

Ayyalah
Traditional dance troupes
Genre: Folk dance

Entropy
Empty Yard Experiment
Genre: Alt rock

Al-Liwwa
Traditional dance troupes
Genre: Folk dance

Bel Bont El3areedh
Hussain Al Jassmi
Genre: Arabic pop

Talaa Al Badr Alaina
Ahlam
Genre: Classical Arabic

Yemen

Read List

The Hostage
by Zayd Mutee Dammaj (1984, trans 1994)
Regarded as a Yemeni classic, this short, 1940s-set novel about a boy held hostage by the local governor to ensure his clan's loyalty is a window into Yemen's era of Imam-rule.

They Die Strangers: A Novella & Stories from Yemen
by Mohammad Abdul-Wali (1987, trans 2002)
The Yemeni experience of exile, displacement and migration threads through this book of work by diplomat-writer Abdul-Wali. It was only collected together in one volume after his death.

Yemen: Travels in Dictionary Land
by Tim Mackintosh-Smith (1997)
This portrayal of Yemen's people, landscapes and deep history on journeys across the country by long-time Sana'a resident Mackintosh-Smith, is one of the best introductions to Yemen for curious readers.

A Land without Jasmine
by Wajdi Al-Ahdal (2008, trans 2012)
Social critique masquerades as mystery tale in this short novel (more novella) about the constraints for Yemeni females, centred around the disappearance of a young woman called Jasmine.

Hurma
by Ali Al-Muqri (2012, English translation 2015)
Set in Sana'a, this coming-of-age story depicts the fate of the young, unnamed female narrator as she navigates the religious and cultural mores expected of her.

Watch List

A New Day in Old Sana'a
(2005; drama; directed by Bader Ben Hirsi)
This simple love story from Yemeni-British director Hirsi gently touches on Yemen's societal restrictions on young people while providing sumptuous shots of Sana'a.

Karama Has No Walls
(2012; documentary; directed by Sara Ishaq)
This Oscar-nominated short documentary by Yemeni-British filmmaker Isaq was shot on the streets during Yemen's 2011 uprising and concentrates on the stories of the protesters who were killed.

The Mulberry House
(2013; documentary; directed by Sara Ishaq)
In this personal view of 2011's uprising, Ishaq turns her camera on herself and her family in Sana'a as the protest movement gathers force.

Yemen: The Silent War
(2018; documentary; directed by Sufian Abulohom)
A searing, short film about Yemeni refugees living in Djibouti's Markazi refugee camp that cleverly uses hand-drawn animation elements to ram home its powerful message.

10 Days before the Wedding
(2018; drama; directed by Amr Gamal)
As some stability returns to Aden, a young couple attempt to organise their wedding, navigating the after-effects of war that impede their lives.

The old (and now war-torn) city of Sana'a is the setting for *Hurma* by Ali Al-Muqri.

Yemen's Bara dance is its most famous folk dance tradition. The number of dancers and movements involved differ from region to region but the Bara is always performed with daggers and accompanied by drumbeat.

The album *Qat, Coffee and Qambus* is a compilation of forgotten recordings from the 1960s and '70s that introduces listeners to the raw folk-blues style that Yemeni musicians were experimenting with during that era.

Playlist

Wa Seed Ana Lak Min Al Khodan
The Three Kawkabani Brothers
Genre: Folk

Ya Saher Aleainain
Abu Bakir Salem
Genre: Khaleeji

Ya Tair Al Hob
Balqees
Genre: Arabic pop

Bara Dance
Sana'a Band
Genre: Folk dance

Mustaq Ya Sana'a
Fuad Al-Kibsi
Genre: Folk

Ya Mun Dakhal Bahr Al-Hawa
Fatimah Al-Zaelaeyah
Genre: Yemeni Folk-Blues

Sadek El Neia
Karama Mursal
Genre: Classical Arabic

Watan Aljerah
Ahmed Fathi
Genre: Traditional oud

Envision
Yemeknight
Genre: Hip-hop

Ana Al Yamani
Ammar Alazaki
Genre: Arabic pop

Zimbabwe

Read List

This Mournable Body
by Tsitsi Dangarembga (2018)
This acclaimed third novel in Dangarembga's *Tambudzai* trilogy was shortlisted for the 2020 Booker Prize. Here the story's protagonist persists in navigating through life's daily struggles until her efforts are derailed by brutal realities of modern-day Zimbabwe.

Under the Tongue
by Yvonne Vera (1996)
Written by one of the country's most famous authors, this award-winning novel tackles taboo subjects of incest and rape.

Don't Let's Go to the Dogs Tonight
by Alexandra Fuller (2001)
This memoir by Zimbabwean-born Fuller reflects upon her experiences growing up on a farm in a white family in the years before and after independence in 1980.

Harvest of Thorns
by Shimmer Chinodya (1989)
The winner of the 1992 Commonwealth Prize for Literature is a novel about a young man finding his way among the politically charged battlefield of pre-independence Rhodesia during the 1960s.

Bones
by Chenjerai Hove (1988)
The award-winning debut novel by internationally renowned writer Hove, *Bones* is a uniquely written story depicting a father's search for his son, who has become a guerrilla fighter in the Zimbabwean War of Liberation.

Watch List

Mugabe and the White African
(2009; documentary; directed by Lucy Bailey & Andrew Thompson)
Award-winning documentary about a family of white farmers who stand up to Mugabe and the contentious land reforms that forced many Zimbabweans to lose their livelihoods and flee their homeland.

Democrats
(2014; documentary; directed by Camilla Nielsson)
Winner of best documentary at the 2015 Tribeca Film Festival, this Danish-made feature looks into the shambolic 2008 Zimbabwean elections.

Albino
(1976; thriller; directed by Jürgen Goslar)
Set in rural Rhodesia, this German film is about a retired policeman's quest to avenge his wife's death after she was raped and murdered by an albino militia leader named the Whispering Death.

Neria
(1991; drama; directed by Godwin Mawuru)
This story of a widowed woman's struggle in everyday Harare is based on the novel by acclaimed author Tsitsi Dangarembga.

Flame
(1996; drama; directed by Ingrid Sinclair)
Taking place during the Rhodesian Bush War, this controversial film is about the troubled journey of two young girls to join the resistance army to fight against white minority rule.

Oliver Mtukudz and his Black Spirits band chronicle modern Zimbabwe.

Performed by one of Zimbabwe's most famous musicians, this soulful and stripped-back track by Oliver Mtukudzi serves as a poignant piece of social commentary that bemoans the devastating effects of HIV across Africa.

A Zimbabwean music legend, Thomas Mapfumo is both an innovator of Chimurenga popular music and a tireless voice against social injustice. *Corruption* is one of many tracks he's written in protest over Mugabe's controversial rule.

Playlist

Njoka
Mokoomba
Genre: Afropop/Funk

Ruva Rashe
Leonard Dembo & Barura Express
Genre: Sungura

Makorokoto
Four Brothers
Genre: Jit

Todii
Oliver Mtukudzi
Genre: Afro-jazz

Chaminuka
Dumisani Maraire
Genre: Mbira

Corruption
Thomas Mapfumo & The Blacks Unlimited
Genre: Chimurenga music

Mandi
Sungura Boys
Genre: Sungura

Skokiaan
Bulawayo Sweet Rhythms Band
Genre: Afro-jazz

Machira Chete
Charles Charamba
Genre: Gospel

Mugove
Leonard Zhakata
Genre: Sungura

Index

A

B

C

D

U

V

W

X

Y

Z

Lonely Planet's Armchair Explorer
July 2021
Published by Lonely Planet Global Limited
ABN 36 005 607 983
www.lonelyplanet.com
1 2 3 4 5 6 7 8 9 10

Printed in Malaysia
ISBN 978 18386 9448 7

Compiled and written by Anita Isalska, Brendan Sainsbury, Carolyn Bain, Carolyn Heller, Christina Webb, Harmony Difo, Helen Ranger, Isabel Albiston, Isabella Noble, James Bainbridge, James Smart, Jess Lee, Joe Bindloss, Kate Morgan, Kerry Walker, Luke Waterson, Oliver Smith, Orla Thomas, Peter Dragicevich, Regis St Louis, Sarah Reid, Simon Richmond, Stephen Lioy, Tamara Sheward, Trent Holden, Yolanda Zappaterra

General Manager, Publishing Piers Pickard
Commissioning Editor Robin Barton
Editors Lorna Parkes, Polly Thomas
Art Director Daniel Di Paolo
Layout Designer Jo Dovey
Print Production Nigel Longuet

Lonely Planet Global Limited
Digital Depot, Roe Lane (off Thomas St),
Digital Hub, Dublin 8, D08 TCV4, Ireland

STAY IN TOUCH
lonelyplanet.com/contact

Paper in this book is certified against the Forest Stewardship Council™ standards. FSC™ promotes environmentally responsible, socially beneficial and economically viable management of the world's forests.